TIME BEFORE GOD

TIME BEFORE GOD

how the universe was created

SHAHIN TAHMASEBI

2nd Edition 2017, pbk.
ISBN: 978-1-925516-98-2
Publishing services by: Greenslade Creations
Designed by: Sharnai James-McGovern

**GREEN
SLADE**
CREATIONS

Acknowledgements

Most of the information on galaxies, stars, planets, and many others about the characteristics, size, and distance are gathered from on-line encyclopaedias, NASA, CERN, ESA, other agencies, and science websites. I acknowledge their hard work and contributions to this book and the greater community. We all wish a very long, healthy, and prosperous life for the Hubble Telescope.

TABLE OF CONTENTS

TABLE OF FIGURES

INTRODUCTION

I have always believed the best business is the business of selling God. I'm not here to sell God, in fact, this book ends when the Big Bang starts. Not many books tell you how they end before they begin. I am not going to discuss the existence or denial of God in this book. Although anyone may have an opinion how the God was created and how he created the universe; but what I'm going to tell you is about the time long long... long before the Big Bang.

When I was a child, every time I asked where everything comes from, the answer was always the same: "God created it", and it was forbidden to ask where God came from, it was a sin. Religion always stops those questions about the creation and God. It is the trade secret they don't want anyone to find out.

In this book, I am going to tell you when time started, when the first proton was created, when, for the first time, light was created, and, most importantly, what caused the Big Bang. You are going to have answers to many questions that have been bothering people for hundreds of thousands of years. These questions have also bothered me a lot, and now I am sharing some the results of my 30 years of research with you.

There are a lot of books and literature that explain how the Big Bang started the creation of the universe, but none that tells about the events that led to the Big Bang and what caused it. One of the encyclopaedias on the web explains the Big Bang as:

> "The major premise of the Big Bang theory is
> that the universe was once in an extremely hot
> and dense state that expanded rapidly."

How did it become hot and dense, and start expanding so rapidly?

If we rewind the Big Bang and put all the planets and stars in the universe back together, it becomes a very large entity, which requires a lot of energy to make it hot to the point of exploding. It would also have required a lot of time for heat to take effect. How did it happen? Where did the heat come from? What were the planets and stars in the first place? Heat is energy and energy must have a source. How was the source of this energy created? You don't believe the heat was generated in less than a second, super-heated the giant entity that existed before the explosion, and tore it apart. Energy generated the heat, and the energy is the result of subatomic activities. If atomic components and particles existed before the Big Bang, then the Big Bang wasn't the start of time. If subatomic activities were present before the Big Bang, then we have time. Doesn't the Atomic Clock work on that principle?

An Atomic Clock works by measuring the microwave frequencies generated by subatomic activities. If heat existed, energy existed; and if energy existed so did atoms, protons, electrons, microwaves, and many other things. The Big Bang did not create protons and electrons: they existed long before. The Big Bang was only the beginning of a series of events that resulted to what we see today, but it wasn't the beginning of creation.

You put explosives in a house and blow it up. You will find bits and pieces of the house hundreds of meters away. A similar process happened to the universe: before the explosion, it was one giant piece of star.

Where did the giant star come from and what caused the explosion? Or, was it only one giant star? Scientists say the Big Bang happened

13.7 billion years ago and that is the time the universe started to form in today's shape. You will soon find out that there were quite a few non-starters—minor Big Bangs—before the main one, and time had started many times and stopped.

If you go to a large event like the Olympics, it usually takes two weeks for the events to be completed. To a visitor, it is only a 2-week event. To a participant, it is 3–4 years' preparation and a few minutes to hours of performance. To the organisers, it is four years of bidding and seven years of preparations, eleven years at least.

With a much larger event, such as the creation of the universe, scientists have always looked at the completing part of the event, how fast the stars are moving and how fast the universe is expanding, but not the creation of those stars. They believe the first star was created sometime after the Big Bang. Here you will find how the stars existed a long time before the Big Bang.

Here, we are going to look at the time before the main event, and what happened long before the big explosion. The 13.7 billion years after the Big Bang seems very short compared to what happened before the Big Bang.

Figure 1: Timeline: the Age of the Universe

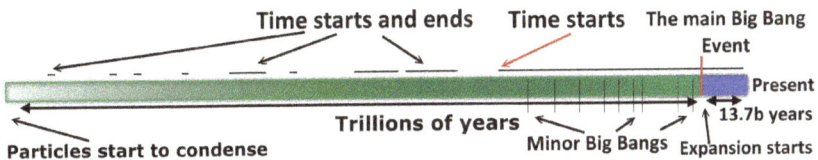

Time starts and ends　　Time starts　　The main Big Bang Event

Present

Trillions of years

13.7b years

Particles start to condense

Minor Big Bangs　Expansion starts

If we were able to rewind the events that have taken place in the universe and go to 1 second before the big explosion, we would be able to see a very, very large entity appearing in the dark. Was it only one entity that exploded into bits and pieces and created the universe as we see it now? Our Galaxy, the Milky Way, contains 200–400 billion stars and many more planets; there are many more Galaxies like ours.

I am going to explain the creation of the Universe from trillions of years ago to one second after the Big Bang. What happens after the first second and after the Big Bang has been well explained; there is a lot of information about it. The easiest way to find what happened after the Big Bang is to stick your head out of the window one night and draw your own conclusions; make sure the sky is clear.

This book starts from Chapter 7; the first six chapters should detail the creation of particles and events trillions of years before Chapter 7 starts. At this stage, no one knows how the particles were created; everyone calls them "God's Particles". Something started the creation of these particles. Similar to the volcanos in Iceland with unpronounceable names (Eyjafjallajokull in 2010 and Grimsvotn in 2011) that pushed a lot of ashes into the atmosphere, something started pushing the particles into space for a very long time, for trillions of years. What it was, if it is still active, and how it could create such small particles we don't know. I hope someone will come up with the first six chapters and explain what caused the eruption. I'm not trying to go off-track and say a volcano-like eruption started the universe; then I would have to explain where the volcano was and on which planet it was located.

This book is meant to challenge some of the beliefs that are now considered the very foundations of science. The readers may find themselves in disagreement with some or all of the areas discussed here; the aim is to engage a conversation and a new challenge.

As you can see in the next diagram, produced by NASA, the first star was created about 400 million years after the Big Bang.

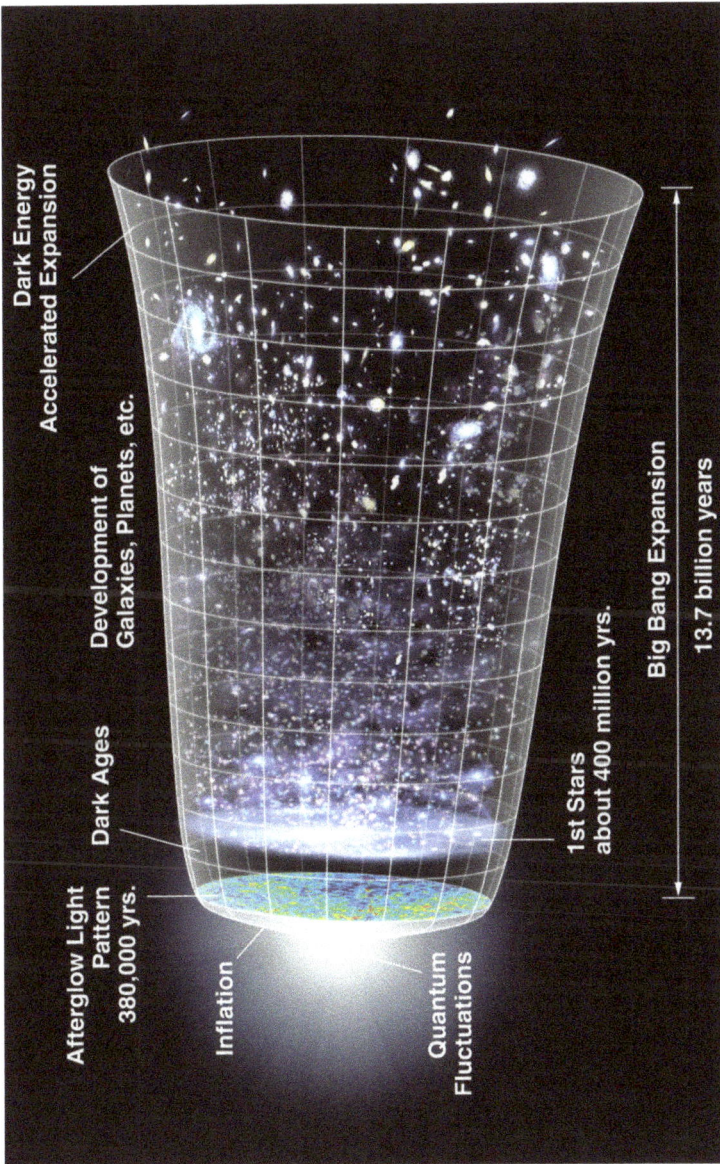

Figure 2: Expansion After the Big Bang

NASA/WMAP Science Team

CHAPTER 7:
IN THE BEGINNING

Time

Trillions of years ago, when time did not exist, the universe started forming. The Webster dictionary defines Time as:

- *The measured or measurable period during which an action, process, or condition exists or continues: duration,* or
- *A non-spatial continuum that is measured in terms of events which succeed one another from past through present to future*

Time is the unit of measurement, and it has to be constant. You can't measure the time travelled between Sydney and London, based on how many times waves smash against the rocks at Bondi Beach. The reference to the measurement has to be constant and accurate.

There are many ways to calculate time. The most popular is the Earth's rotation on its axis and around the Sun; the Moon's rotation around the Earth is also used. These rotations define the year and the year is broken into days and hours to give an accurate reading of time.

One of the most commonly used tools to calculate time is Quartz. Quartz is a crystal found in nature that generates signals with a very accurate frequency; the frequency is 32,768Hz. This accuracy is commonly used by watchmakers to create accurate time counters (watches and clocks).

Scientists have calculated the beginning of time to 10^{-43} of a second from when the Big Bang started; it is a zero, a point, and 43 zeros after that and then a 1. There is no need to print all the 43 zeros here. That means there was a bright white light and in less than one second, (my public education doesn't allow me to count anything that has more than 15 zeros) the explosions start and the universe starts to form. Stars appear, galaxies are created, and so on. What happens after this does not concern us here; everyone knows what happened.

Scientists say time started from here. You will learn that time started billions and billions of years before that. The Big Bang was the start of a new chapter in the history of the universe, but it wasn't the beginning of it.

We do not know exactly when time started because there were quite a few false starts and a few minor Big Bangs. However, it certainly started long before the 10^{-43} of a second after the big bang as has been stated; in fact, billions of years before.

To get to time and its starting point we need to visit some of the other areas first. Points like SPACE, MASS, ENERGY, LIGHT, VELOCITY, SPEED, and my favourite ANGLE. Why angle? Angle was the first law of geometry. The first law of physics was the Movement. Something that actually moved was the first in the universe.

The use of the term "years" in this discussion is wrong because it refers to an event that has not even begun. It measures the time in units that were not yet created. The year is how long it takes the Earth to circle the Sun once.

The time discussed in this book is well before the creation of the Earth, Sun, and of course the Universe. To give a sense of understanding to the readers, I have no choice but to continue using this term, "years", just to convey the concept of time. In order to understand what the book is talking about; we need a neutral reference to base our understandings. This is a concept of reference, as a bridge, to assist in understanding the events.

In fact, measuring long distance units based on Earth's rotation around the Sun is not ideal, but we have no other means of doing so.

Space

This section explains how the laws of physics were created first. To get to that point, we will consider how Space started. This book starts, not exactly from the beginning of when particles were created, but a very long time after the creation of particles. The events registered here are maybe less than one percent of the time since the first particle was created. So if these events started trillions of years ago the particles' creation started a sextillion (10^{21}) years before that, at least. How the particles were created, we don't know, yet. The definition of Space according to the Oxford Dictionary is:

A continuous area or expanse, which is free, available, or unoccupied

If space is made out of nothing, it is an empty area in three dimensions of X, Y, and Z. Space is made out of nothing; it doesn't even exist. Well, maybe, but no one in the whole universe can create it. It is like creating nothing, creating an empty area; but it already is there. On the other hand, it is also correct to say everyone can create "nothing"; even newborn babies create nothing; well, besides the other stuff newborns do.

Figure 3: The Size of the Universe

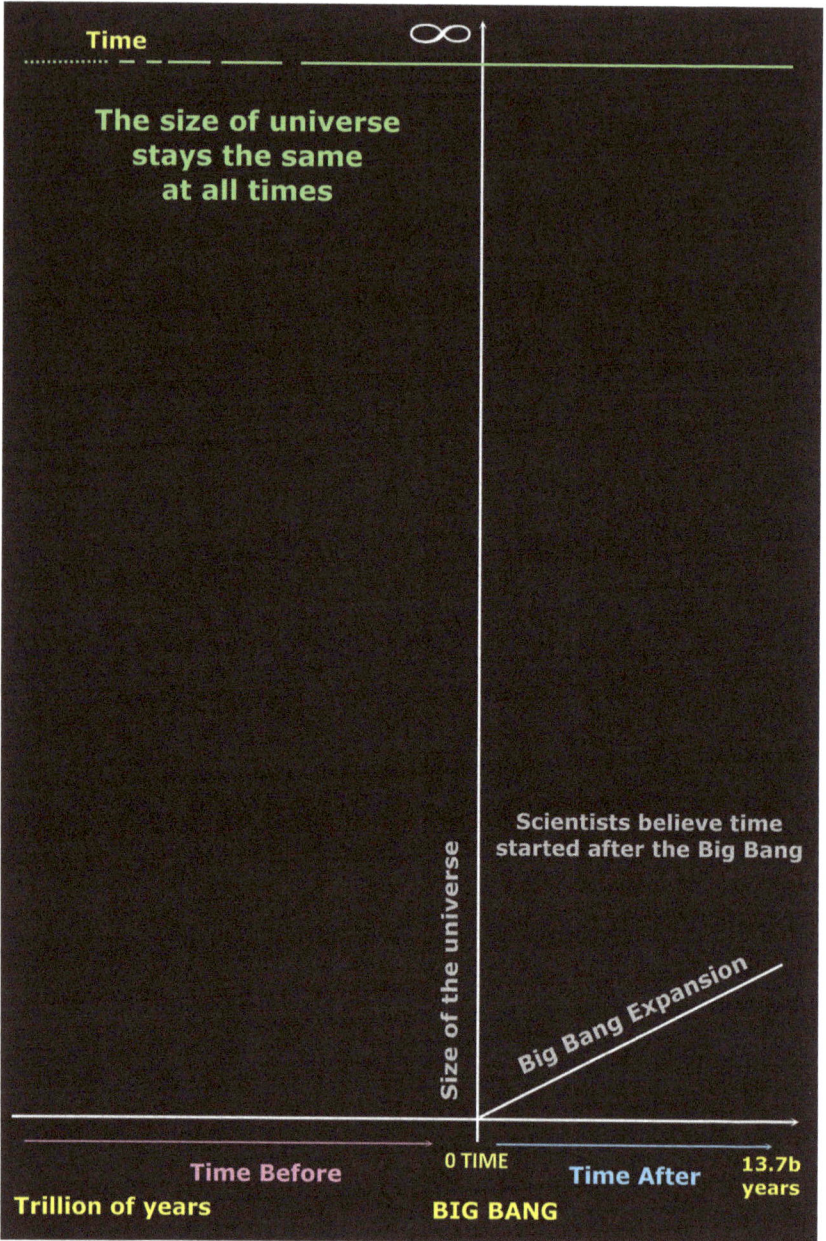

Time ∞

The size of universe
stays the same
at all times

Scientists believe time
started after the Big Bang

Big Bang Expansion

Size of the universe

0 TIME

Time Before

Time After

13.7b
years

Trillion of years

BIG BANG

Creating "nothing" doesn't need any energy or qualifications; anyone can create it. So, if no one created space, how was space created? It wasn't; no one creates the "nothing", space is not an entity, and therefore no one can create it. Space is an empty place that is occupied by some planets and stars, ah! ... and some asteroids too.

Space cannot have borders: it cannot be confined to enclosed areas. Once you put a border around the space, then you've lost the definition of space. Emptiness and Nothingness do not exist as entities, no one can create it, and no one can claim ownership over it. Space is only a name we use to call the empty place out there. You can say empty room, empty box, empty house, empty building – but not empty on its own. Empty is an adjective. The definition of EMPTY, according to Oxford Dictionary is: *Containing nothing; not filled or occupied.*

If space is filled with nothing and contains nothing, it cannot exist as an entity. Since it doesn't exist (as an entity) it cannot be moved, relocated, bordered, contained, be drawn in a picture, be photographed, and most importantly, it cannot be bent. To bend space, you need to convert it into an entity. Nothingness cannot be converted to an entity and cannot be bent. So, if you are trying to travel from A to B and thinking the shortest way is to bend the space, first, you need to come up with a solution to convert space to an entity. Space is at the current shape as it is now, nothingness, and emptiness cannot be bent unless it is turned into an entity. Very simple: you go and try to bend the 10 inch PVC pipe that you don't have, let's see if you can create a 45° bend.

Scientists have estimated the diameter of the observable universe at approximately 93b light-years; it doesn't mean space ends after 93b light-years. If we had a ship and were able to travel that far, what would we see when we reach the end of this journey? Would we hit a wall? Is there a drop? Or, maybe it is a round shape, and we will end up where we started.

The size of the universe is infinite and has no end.

Temperature

In a very, very far distance in time before the Big Bang, time didn't exist; space was just an empty place. Temperature didn't exist; it was thousands of degrees below Absolute Zero, and it was so cold that if you had an entity and threw it in there, it would disappear immediately. In such a cold temperature, subatomic activities cease to exist. Atomic activity creates complex molecules and therefore an entity. If the atomic activities cease, the molecules fail to form, and the entity falls apart and disappears, as atomic particles are not visible to the naked eye. What holds any entity together, whether it is a living organism or any other entity, is the atomic activities that create the molecules and, thus, the entity.

Absolute Zero is −273.15°C; at around −250°C hydrogen is still solid. At −252°C and extreme high pressure of around 400GPa (14,503,263 psi), hydrogen becomes a form of liquid (superfluid) consisting of protons and electrons that are at Zero-Point Energy (ZPE). It means that electrons are not spinning around the protons and atoms are not formed anymore. Protons and electrons form a pool that is in the form of liquid that could be poured from one container into another like water, but they do not bind to each other. It is like a cup of sand with the grains next to each other but not binding.

Absolute zero is the theoretical temperature at which entropy reaches its minimum value, and Zero-point energy is the lowest possible energy that a quantum, mechanical, physical system may have.

The ZPE is not quite zero-point energy as scientists have told us.

Figure 4: Structure of a Proton

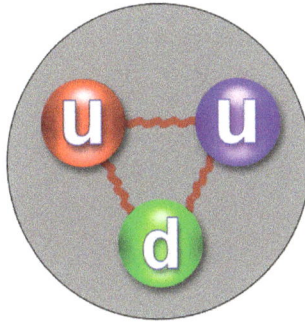

Protons are made of smaller particles called quarks. Quarks are the most elementary particles that make up matter, and have mass, occupy space, and have fractional electric charge values. Protons are made of three quarks, two ups and one down, the scientists believe. At absolute zero, sub-atomic activities cease but not the activities of the quarks inside the protons that keep the proton together. To break a proton, the temperature would have to go well below absolute zero and the pressure would need to go much higher.

One thing I don't understand is, if the protons are like a hollow sphere with three quarks exchanging electric charge inside, then it means the particles have an excess of electrons. I thought "electric charge" is the physical property of the particles and when did the particles inside the proton become electrically charged? Electric charges are of negative and positive types, and that requires electrons and electrons are larger in size than the particles. How did the particles get inside the sphere, and how did the sphere close on them? Where was the electromagnetic field to charge the particles? Electromagnetism requires a source, and, in this case, it requires a giant source and the source needs energy. If this is the case, then what is the wall of the sphere made of? It doesn't sound right, and I will explore this later in the chapter when the creation of the first proton is discussed.

The electric charge is the elementary electric charge of particles, (denoted as *e*). The electric charge of a proton is positive, and the symbol is: P⁺ or Electric Charge: +1*e*.

Figure 5: Metallic Hydrogen Temperature to Pressure Ratio

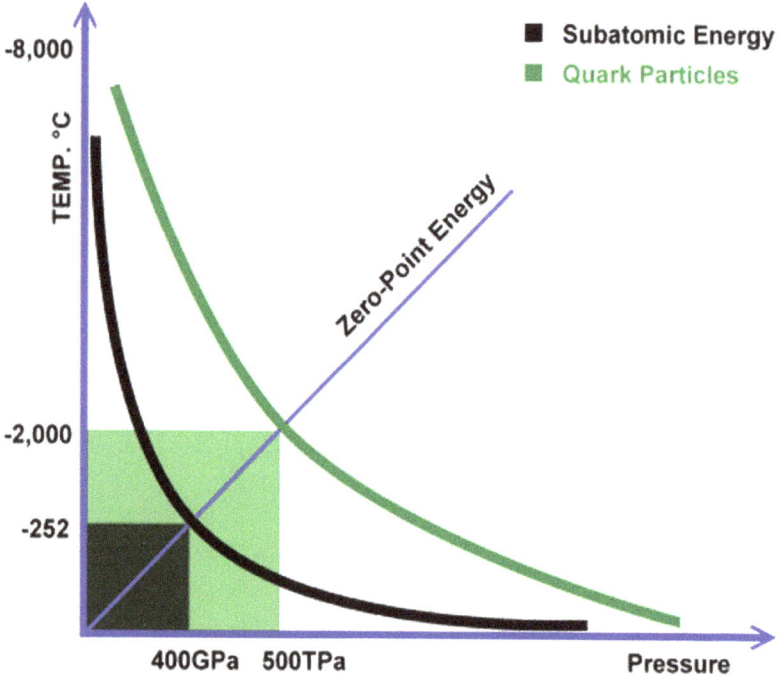

If the temperature is at the levels that existed trillions of years ago, molecules cannot remain together, atomic activities cease, and subatomic particles collapse. Theoretically, you can smash the electrons and protons into pieces. Of course, the air and atmosphere don't exist as they, too, are frozen, and their atomic structure has collapsed, so they have disappeared. At that temperature, energy in any form does not exist; not even in the lowest possible form.

Figure 6: The Boomerang Nebula   Figure 7: The Boomerang Nebula

NASA/ESA

NASA/ESA

The coldest place scientists have found so far in the universe is the Boomerang Nebula located 5,000 light-years away from Earth. With a temperature of −272°C, "...*contains ultra-cold gas at temperatures below the microwave background temperature, making the Boomerang Nebula the coldest place in the universe found so far*". At this temperature, molecules are not formed anymore, subatomic activities cease to exist, and the atmosphere is not gas anymore: it would be in a metallic state with Zero-Point Energy, but scientists say it is a gas.

The clouds around the dying star either are not gas but are a metallic vapour of what used to be gas, or are still gas even though the temperature is at almost absolute zero. If the cloud is a metallic vapour at Zero-Point Energy, protons and electrons are not in a binding state at −272°C, then Kelvin's theory is correct, and temperature cannot go lower than −273°C. But the photos taken by the Hubble telescope in 1995 show the cloud can reflect and fracture light (the light in the photo is from the stars behind).

Light is fractured as it passes through molecules when the medium composed of those molecules changes. Light changes direction, intensity, and frequency when it passes through molecules of a medium of a different density, like travelling through air and then water. If light goes through the gas cloud and the frequency has changed then, the molecules create the cloud, as scientists have correctly identified. However, if the temperature is −272°C, then why are the molecules not in at least solid state and Zero-Point Energy (ZPE)? The light should not fracture and change its frequency when the medium is at Zero-Point Energy and molecules are not formed. This proves that, in the absence of extreme high pressure, the temperature needs to be a lot lower to stop subatomic activities. At ZPE, gases don't exist, anymore, as we know. There will be no molecules that form the "gas cloud", therefore there will be no change of medium; it will be all the same medium: empty space.

If a light beam travels through air and hits clouds, and water vapour floating in the air, where the light beam enters the water molecules, the medium is changed, light fractures, and the frequency of the light wave changes; it creates a spectrum or a rainbow. If we place the same water vapour in space far from any planet or stars and reduce the temperature to −272°C, water molecules go to ZPE state and disappear. What will be left are the protons and electrons in a metallic state that used to be hydrogen and oxygen. If it were a thick cloud that we created, it would become a thin layer of cloud that is completely black and doesn't let the light pass through. A light beam entering it is stopped, like hitting coal grains; no light rays bounce off the protons and electrons as there is no shiny surface to reflect, and, because the low-temperature energy of the photons tends to be absorbed by the protons, light is not fractured in this cloud of particles but is totally absorbed.

If there were a large cloud-like entity made of particles at ZPE, it would be seen from far away as a very black entity that does not allow light to pass through. At −272°C, a cloud of any gas should become a cloud of dust similar to dust storms we experience from time to time. The Boomerang Nebula should appear as a black entity that doesn't pass any light through as, at ZPE, particles block all of the light rays.

Figure 8: Boomerang Nebula at −272 °C (left), and at ZPE it Should Look Black (right)

Images from the Boomerang Nebula show temperature alone is not enough to reach ZPE. Pressure is also a contributing factor. It also shows that to reach a ZPE without pressure in a vacuum; the environmental temperature has to go a lot below −273°C. Putting a limit on temperature in the natural environment is not right, whether at high temperatures or low temperatures; even if those temperatures cannot be reached, it is wrong. We haven't seen and haven't been in every place in the universe and assuming temperatures cannot go lower or higher than a certain point is wrong. Until we go everywhere in the universe, test the environment, get a reading, and observe the condition of the subatomic activities, we should not presume there are limits on such events.

It is like a fire brigade evacuating a large shopping centre: they are not going to say the centre is evacuated by checking the first ten shops and rooms. The only way they can announce a complete evacuation is to search the centre room-by-room, door-by-door. Because we tested in the laboratory and could not reach temperatures of lower than −272°C, it does not mean it cannot happen anywhere else in the universe. There are places in the universe, trillions of light-years away, that are still in the original state before the condensation begins, trillions of years before the Big Bang.

The image below shows how space should appear from a distance with a cloud of particles at ZPE. The pictures taken by the Hubble Telescope outside the Earth's atmosphere are extraordinarily clear and sharp; they show no presence of ZPE clouds of particles. Of course, we haven't looked at all places in the universe, but where we have looked, seems not to have these clouds or the scientists have overlooked and ignored the importance of such clouds. I don't believe these clouds exist anymore, over trillions of years they have dispersed, burnt, or just disappeared, falling into planets and becoming active again as molecules after a rise in temperature.

Clouds of Particles

ZPE clouds of particles are in two shapes: Deformed and circular. The uncommon, deformed, shape of cloud, like we see on the horizon on Earth, is stationary and not moving at all. The circular shape, like black holes we've seen in pictures from the Hubble telescope, has a large number of stars and planets circling around it.

These ZPE clouds are made out of two different particles. They are both black because light cannot escape from them.

The first instance (Type-1), cloud-like shape, is made out of electrons and protons at ZPE. They used to be molecules and had subatomic activities. They were gas clouds ejected from a dying star or leftovers of collision between planets. They became cold after billions of years, and the temperature went down to around −273C, they lost subatomic energy. Protons and electrons became separated from each other and then created the cloud, as we see in some images from deep space.

Figure 9: Cloud of Particles at ZPE

The second form of ZPE cloud (Type-0) is made out of the original particles before the condensation began. Either these particles did not have enough time to join the rest of the particles to form the universe, or they were too far from others to be part of the condensation (I will explain shortly how these particles were in space trillions of years ago and what the condensation was and how it started). These particles, a very large number of them like an ocean, remained after the Big Bang explosions, and the energy that was created by the Big Bang pushed them like a hurricane and started them spinning around their axes. Because there is no atmosphere in space and no resisting force, they can spin forever. There are two different types of particle, Type-0 (smallest particles) and Type-1 (larger particles made out of protons and electrons).

The difference between these two particles is that Type-1 particles are made out of protons and electrons at ZPE, and can go back to the original state they were in before. If the temperature rises for some reason, they can go back to what they were before, form molecules, and have their subatomic energy back. If there is an asteroid crossing their path, it can also push some of them out of that area, and they will eventually become warmer and become active. If the size of this cloud is very large, it can neutralise the asteroid or the planet's subatomic and molecular activities and turn the body into a ZPE cloud. This event will help the size increase of the ZPE clouds very little because the planets and asteroids are warmer, and the clouds will eventually become warm and will turn into clouds of cold gases.

A planet like Earth can cross into such clouds, and disintegrate very quickly if the ZPE clouds are very large. If the size of the ZPE cloud is small, the planet will start to lose its atmosphere as it enters the cloud. The molecules that create the atmosphere are the first to become very cold, very quickly, and gravity pulls them down until the cold hits the surface of the planet and turns it into particles. This process continues very fast until the temperature balances; the cloud becomes warm as it takes the heat from the planet. The process continues until the entire planet is frozen. It may not turn into a ZPE cloud, but it will become a dust cloud.

It is different with the second form of ZPE clouds, Type-0. These clouds are formed from the particles that are smaller than protons and electrons. They are the original particles that were not turned into subatomic particles. Temperature has no effect on their formation, as the current temperature in space and lack of pressure will not be enough to turn the particles into subatomic particles, as happened after the condensation began.

A planet or asteroid can easily enter into this cloud, and be turned instantly into Type-1 particles. Type-1 particles are larger than Type-0 particles. Hypothetically, you can collect Type-1 particles in a container, the molecules of the container will stop the particles from falling out; but Type-0 particles are smaller than any molecules, and they go through the container and fall out, destroying the container molecules in the process and turning it into Type-1 particles.

Figure 10: Size Comparison of a Proton and a Type-0 Particle

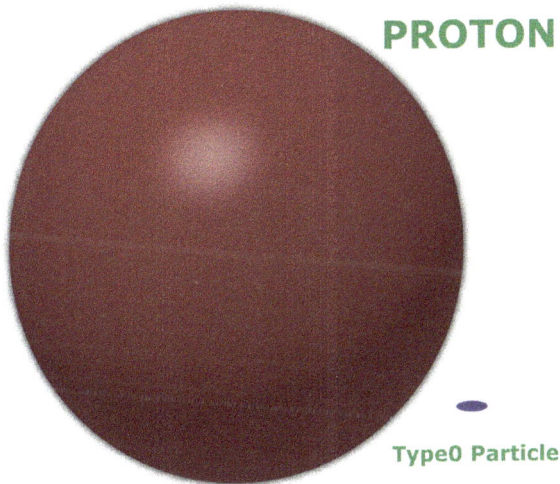

PROTON

Type0 Particle

The shockwaves from the Big Bang pushed large amounts of Type-0 particles and started to spin them around their axis. For billions of years after the Big Bang, they continued to gather forces, consume planets, and increase their size. By consuming planets and turning those into Type-1 particles, what was left of them—protons and electrons—were pushed to the outer edge of the whirlpool clouds.

Uranium (U238) is used to enrich Uranium in nuclear plants. The Uranium is turned to gas using acids and other catalysts and then is pushed to a centrifugal processing unit. This unit is a machine similar to a clothes dryer that is spinning very fast. Strong centrifugal force pushes the heavier isotopes outside the filtered zone, and they are collected in a separate process. The condensed product that contains heavier isotopes is called Uranium Oxide (U308) or Yellowcake. This process is long and complicated.

Figure 11: Yellowcake

US Government

The protons and electrons are heavier than Type-0 particles; unlike the uranium enrichment, it is a very slow but long spinning process. After millions of years, the heavy protons and electrons make their way through the Type-0 particles to end up on the outer edge of the whirlpool Type-0 clouds. This very interesting process happens over hundreds of millions of years. The Type-0 cloud is there like a large disc in space, a planet, large or small, slowly comes closer and closer to the cloud. It enters the Crust or outer belt of Type-1 particles, then the particle cloud, the Type-0 particles start entering the molecules of the planet. The molecules slowly start to disintegrate. The molecules disintegrate first, and then it comes to the subatomic particles.

It takes longer for the subatomic particles to disintegrate. The Type-0 particles go between the protons and the electrons causing them to stop spinning. It slowly gets harder and harder for the subatomic particle to stay in formation. Slowly, the deeper the planet enters inside the cloud, the more the planet disintegrates. This process is a very slow occurring event; it takes millions of years to complete.

The disintegrated molecules and subatomic particles that formed the planet are slowly gathered around the outer edge of the clouds like a crust. They slowly start to form new molecules but will not form a new planet, just small bits and pieces. They don't even have a chance to form large entities as they are constantly invaded by the Type-0 particles and disintegrated again. This process of joining and separating of the subatomic particles continues to happen almost forever. Strong energies are released from the planet while it is disintegrated. When the destruction closes the core, a sudden release of energy causes a pulse and shockwave across the disc. Although the sudden release of energy by the destroyed planet is powerful, it is very small compared to the size of the whirlpool disc.

The new small bits and pieces that have been created in the crust cannot bind and create larger entities as no catalyst, like heat, which helps to create new molecules exists. The Crust is near absolute zero, so cold that hardly any molecules can stay in formation and binding. The consumption of the planet by Type-0 particles reduces the planet to a ZPE Type-1 cloud. The size of such cloud of particles is so large that consuming a large planet only adds a little to its size.

This process of consuming planets increases the total size of the cloud of particles. Type-0 particles are in the centre and Type-1 particles at the outer edge; further away is the crust. Light cannot escape from this cloud and the crust as they all absorb the energy of the photons.

Temperature has played and still continues to play, a vital role in the creation of the universe. I believe, if there is going to be the first law of physics that started the universe, it is temperature. Maybe there were a lot of particles in space, but without temperature, they would still be particles as they were for trillions of years before the condensation began. Maybe some argue the first law of physics was motion: particles moved to a place that later started the creation or the creation of the particles on its own was the first law of physics. None of that could have transformed the ocean of particles into the universe as we see it today without temperature. First, the low temperature brought the particles together and then the high temperature resulted in the Big Bang and the rest. Still, trillions of years later, temperature plays a vital role.

The large volume of Type-0 particles, long before anything started, was stretching hundreds of billions of light-years across empty space. If you could be in a spaceship and fly through the particles, you could have the particles enter the skin of the spaceship and pass through your body without any problem. It looks like the ship, your body and anything else is made out of mosquito net. The particles are so small that they can pass through the molecules. This can cause molecular formation and subatomic activities to cease.

Let's use our imagination and find a spaceship that has a body made of a material with a molecular formation so small that it can stop Type-0 particles from passing through. This material would be made out of something other than the elements we can find in the Periodic Table: something that has the space between its protons and electrons, smaller than Type-0 particles. The molecules made from these elements also have the smallest space between them. This material can be used to create the outer skin of the spaceship; the rest inside is made out of the stuff we have now.

The spaceship is just outside the ocean of particles. It is hovering about a mile away and is facing the ocean. If you look at the pictures taken of sandstorms, everything looks normal just before the storm hits. Then it becomes black.

Figure 12: Dust Storm: Mungerannie, South Australia

The spaceship sits a mile outside the ocean, and we are looking directly at the particles. Behind us, about 10^{24} light-years away, a new universe is about to begin. Type-0 particles have started to form; a new creation has just begun. The very first moments of creation have started. Time and distance have no meaning. Distance is so great and time is so small that it cannot be measured, or if it could, it would have no meaningful result; they are just bunch of numbers that we struggle to consume. It is possible, not definite, that the start of the creation of another universe in the very far distance would begin. There are many factors that have to come together to start a universe. We should not assume many more universes like ours exist. One has already been created, ours, and another one is being created, but not many more.

Ahead of us, we can't see anything because it is pitch black. There is no natural light emitted; no stars are yet created. The only light source is the lights from the spaceship. We turn them on and still see nothing. The light beam from the ship reaches the particles and is immediately and completely absorbed by them. Nothing is reflected back. We see entities because of reflection of light beams from the entity, or direct emission from the source. We can see the coffee table in the lounge room because of the redirected light beams from the bulb hanging from the ceiling, and we can see the bulb from the beams emitted from it.

No matter how powerful the light source of the spaceship is, we cannot see anything, not even one tiny ray of light is reflected back to us. We don't even know if we have reached the particles yet. If we could see the particles we could see a very large lump of black stuff that stretched for eternity, we can't even see the end of it; it goes forever. I can't even put a figure on it.

The Particles Start to Move

Our journey starts just outside the particles; the time is about 70–100 trillion years before the Big Bang. The condensation phase is about to start. These particles have been here for trillions of years since their creation. Now so cold, they have started to move towards the centre and closer to each other. With the instruments on board the ship, only a mile away, you can see the ocean is moving very slowly. It is contracting. The rate of movement is so small that it cannot be verified by anything we now have; it has to be a very special tool. The contraction is less than a meter in every thousand years.

The particles don't respond to light, so the use of a laser beam to measure distance is not possible. Microwave signals move them slightly, but these do not come back either. The only way to make the particles move is to generate a powerful low-frequency signal, something like 1–4Hz. This signal, if generated very close to the particles can move them. This signal is not going to penetrate deep inside and travel too far. It moves the outer skin, is absorbed very quickly, and disappears.

A Hz signal will not travel in space; it travels in the air here on earth but not in space. The signal generator needs to be very close to the particles and the particles act like molecules in the air and allow the signal to travel. Of course, this entire project is done "assuming" we can generate such a signal that the particles do enter the molecules and subatomic particles of the device and stop them from binding.

The shape of the mass of particles is interesting. A drop of water in the air, or space, is a sphere, a perfect round shape due to the equal forces pushing out from inside making it spherical. At this time, before the universe was created, there was no energy, and therefore the shape of the particle mass would not be a sphere, it also is not made of molecules that are connected in formation like a liquid. If we could see, it looked as though it had no particular shape, similar to the dust cloud it was all over. The start of the mass is made of very low-density particles and gets a little denser deeper inside but not very dense.

The spaceship enters the clouds; the outer skin is made out of special material with molecules that have no empty space between them; we shouldn't, in fact, call them molecules. Similar to an aeroplane that pushes the air this space ship pushes the particles. The thin outer skin is made out of this material, a special paint like film; the rest of the ship inside is made out of steel and aluminium. It is possible to manufacture this material. Everything in the universe is made out of molecules, some very simple and some very complex. Molecules have empty space between the atoms and subatomic particles. The Type-0 particles are smaller than these empty spaces and can pass through and become lodged between the atoms, preventing molecules from binding.

In a long and difficult process, similar to the production of Yellowcake, protons can be separated from atoms of different metals like lead, collected in a powder form, and placed between two sheets of thin metals. This powder will work as a barrier. Type-0 particles pass through the first metal sheet, going through the molecules, and are stopped at the protons, as there are no spaces between them. The space between the protons is so small that even if some particles still get through, they will stop eventually. This shield will not stay intact for long as the particles will turn the first metal sheet into small particles of protons and electrons dispersing them after a while and the shield will then collapse completely.

I don't want to go into the details of the ship and how it is made and the instruments in it, otherwise it makes it like a science fiction book. I'm not trying to write a science fiction book or tell a story; I tell it as it happened. Let's say it is not even a spaceship; it is an armchair. We are sitting in a comfortable armchair and travelling through the cloud of particles. Temperature is about −8,000°C and the time is about 100 trillion years before the Big Bang, the 13–14 billion years after the Big Bang is too insignificant to bring into account. Here we are between the time the particles were created and the Big Bang. How and when the particles were created is something that is not discussed here. In recent times, scientists have said that evidence can be found that shows the Dark Energy exists and the Dark Energy started the Big Bang and therefore the universe. Where we are now, goes back trillions of years before the existence of Dark Energy. At this time, no energy existed, whatsoever, in any form. The very first energy in any form that was created was at least 50–60 trillion years after the time that we are in, the time on the armchair looking at the very first particles that start to condense.

Stages of Creation

The creation event took place in 3 stages:

- **Stage 1:** Creation of Particles. Timeline: We don't know
- **Stage 2:** Transformation of particles into visible matter. Timeline: 50–100 trillion years
- **Stage 3:** Creation of life. Timeline: 14 billion years

Figure 13: The 3 Stages of Creation

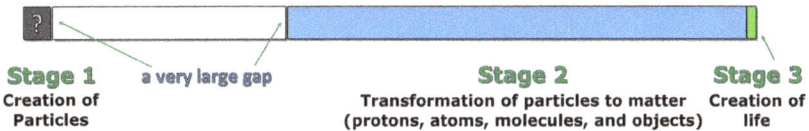

The time has been divided into three stages. First, the time that goes right to the beginning of particles and how they were created; the second stage starts when a very large amount of particles was in space. So much that it looked like an ocean that goes on forever and never ends. The third stage is the time after the Big Bang until now. The first six chapters will eventually explain how the particles were created, and the process that created them. In this chapter, I look at Stage 2 when the particles already existed, and space was filled with a lot of particles. To understand how the particles got there, read the first six chapters! We need to understand nothing happened between the stage 1 and stage 2 for 1000s of trillions of years.

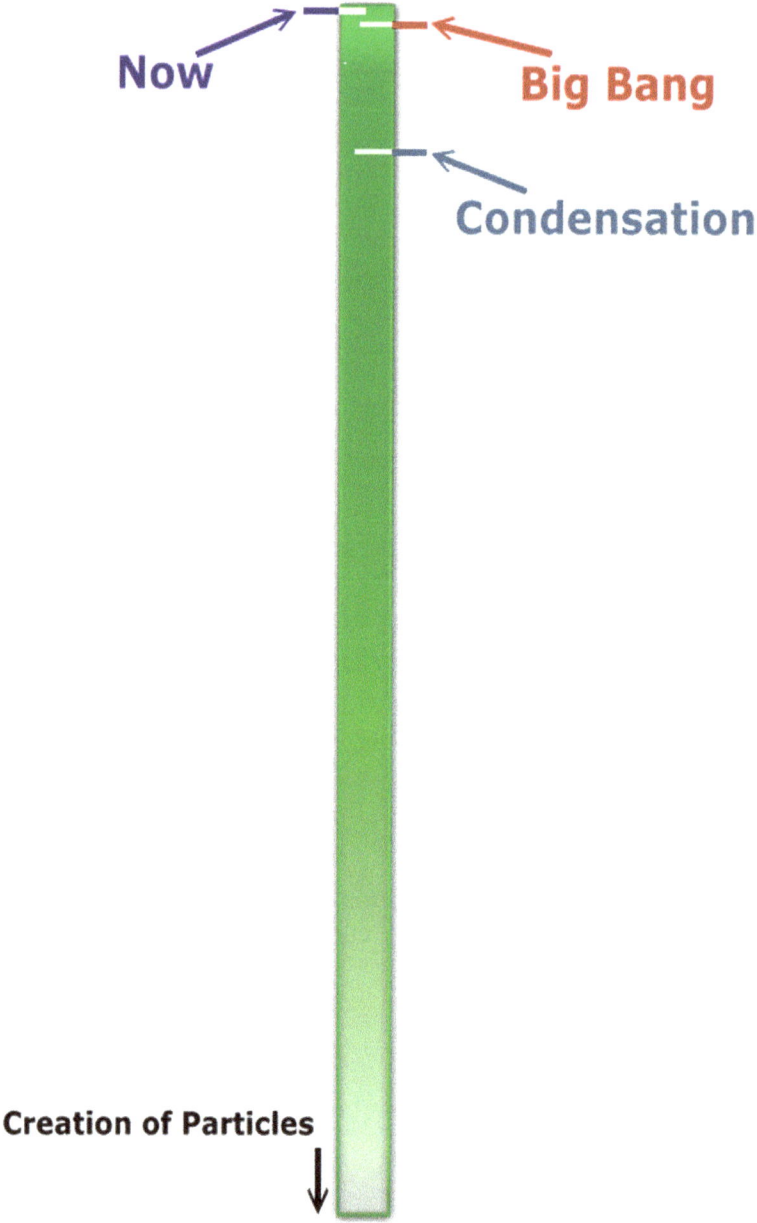

Figure 14: Particles Timeline

Smashing Protons

CERN (European Organization for Nuclear Research), established in 1954, has built one of the world's largest particle laboratories on the border between Switzerland and France. They built the largest Hadron Collider that shoots protons and smashes them into each other at almost the speed of light. Nothing new about this process and the collider, this device existed in a much smaller version since the first television was introduced in 1926.

Figure 15: CMS Detector

© 2008 CERN

In old TVs, the Cathode Ray Tube (CRT) works in a similar way to the Hadron Collider. An electron Gun at the back of the tube shoots electrons towards the screen coated with phosphors at the front. These electrons are directed up and down by an electromagnetic field created by two horizontal and vertical coils called a Yoke. If the Yoke is not active, the electron goes straight from the gun and hits the phosphors in the centre of the screen causing it to glow (the white dot). If the horizontal coil is turned on and the electromagnetic field is created, the electron is directed to the North or South of the field changing its direction and the collision course. This process happens 100 times per second (the electricity cycle is 50Hz). This is the shortest description of how the Particle Collider works, but it remains one on the world's most complicated and most difficult science experiments to date.

The Hadron Collider is a larger scale and very much more complicated device than a television but similar in principle; it shoots protons instead of electrons. It shoots two protons in opposite directions in a vacuum tube and speeds them up to a speed near the speed of light by using very large electromagnetic devices. This process continues for a while until the protons reach the maximum speed then by changing their direction using electromagnetic fields they are put on a collision course and are smashed into each other; a very complicated process but based on simple principles. The scientists working on this project claim the protons are broken into pieces and are the "God's Particle" or the smallest particle that everything else is made of and which started the universe (the Type-0 particles).

Figure 16: Large Hadron Collider

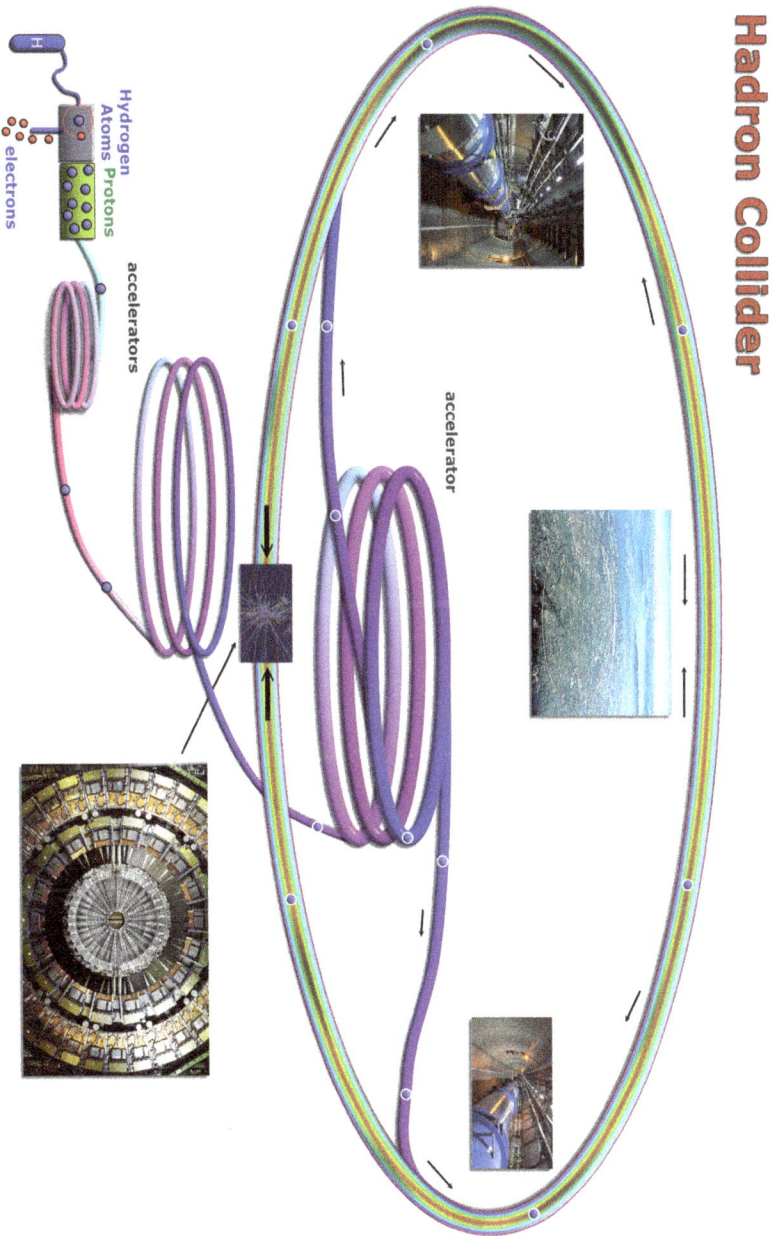

To understand the result of smashing protons, we need to look at black peppercorns.

Pepper or black pepper is a fruit of the Pepper plant. The fruit grow, like vine grapes, in clusters and after picking are left in the sun to dry. The sundried corns are placed in a milling machine and turned into cracked or finely ground pepper.

Figure 17: Black Peppercorn

The colour and the taste of black peppercorn comes from the skin of the fruit; the inside is white. A good quality black pepper powder is very black and more expensive because only the skin is used. The picture shows a black peppercorn on the left and after it has passed through the milling machine on the right. Protons look very similar to the black peppercorn: at the time of creation, they have a very rough surface; soon after, they become very smooth like passing through a milling process.

Similar to the peppercorn, protons can be smashed, crushed, or finely ground. In the future, when technology is advanced, maybe we can develop a machine with counter-rotating disks, like the milling machine used for pepper, to grind protons. This may seem a crazy idea now but, maybe, in 100 years from now, we can develop a paint-like substance made only from protons to give plating to metal disks (similar to chromium plating). Then we could build a small milling machine the size of a large domestic fridge, which could grind protons in a laboratory rather than a 27km long underground tunnel under two countries, and not as expensive.

The proton powder can be used in solving the problem with the greenhouse gases for example. A large filtration plant can be built that holds the powder and pollutant gases can be passed through the filtration where the proton powder can neutralise and break down the gas molecules with a minimal use of energy. From one end CO^2 is pushed through and from the other end protons and electrons are collected in a separate process that can be put back together and have the Oxygen and Carbon back. This is a very cost effective and efficient system to tackle the greenhouse gases. It can also be used to dismantle radioactive molecules. Radioactive waste can be processed in a similar manner to dismantle the radioactive molecules and produce neutralised electrons, protons, and isotopes.

This machine, I call it the De-Moleculeizer, works with the minimal use of energy to break up the molecules' structure and turn them into single atoms or single protons and electrons. Nature is already doing this for us if we let it. The trees, for example, collect the CO^2, break it into Carbon and Oxygen by using sunlight, store the carbon in the trunk, and release the Oxygen into the atmosphere. By passing the poisonous gases through this machine, the proton powder particles fill the empty space at a subatomic level and cause the molecules to break up and dismantle.

Of course, the De-Moleculeizer will give a good excuse to some to turn the old growth forests—if there are any left by then, into coffee tables and wardrobes.

Although the protons are made from the Type-0 particles, smashing a proton will not give us the particles back. The bits and pieces of broken protons are not the Type-0 particles or the "God's Particles". In fact, proton powder in size is closer to the Type-0 particles than smashed bits of proton. Smashing two glass marbles will not give us sand no matter how hard we smash them against each other. We will get some broken bits and pieces of broken marble but not sand. By reversing the process, we will not be reversing the creation. The creation cannot be undone. It is not like a reverse engineering of a car. By putting a car into a shredder, we will not get the bolts and nuts that once held the door hinges.

Figure 18: Smashing Two Glass Marbles Will Not Give us Sand

The properties of the particles changed when they turned into protons and electrons. By smashing protons, we will have bits and pieces of different sizes of broken protons but not the Type-0 particles.

Raindrops are, on average, 1.2mm in diameter (due to atmospheric conditions and gravity); they start at 1.2mm when leaving the clouds and very quickly join other droplets and create larger droplets of sometimes up to 5mm in diameter. The Type-0 particles, when created, had all been very similar in size. Although there was no gravity or atmospheric conditions but under circumstances and conditions of their creation, they were created the same size. (None of the particles is in the shape of a sphere; a droplet of rain in the air is spherical because of forces within pushing out at the same rate, which makes it a sphere. At the time of creation of the protons, there was no energy or force, and it could not have pushed out to make the shape of the particles spherical). The force of impact in the Hadron Collider blows a large part at the back of the proton, possibly two smaller size particles on the sides, and much smaller broken bits at the front. The energy from the collision is transferred to the back of the sphere (Newton's Law of Motion, and Newton's cradle) and blows the back first.

Newton's Third Law of motion:

> For every action, there is an equal and opposite reaction.

Type-0 particles cannot be created in this universe by humans or by the universe itself. The conditions required to create particles do not exist in this universe, and we cannot create those conditions in the lab. We might create particles as small as Type-0 but not the Type-0 itself. The conditions that existed, and still exist in the Greater Universe, which were responsible for the creation of the particles do not exist in our universe anymore.

Another interesting point is that no protons, neutrons, or electrons can be created in this universe; even the universe itself cannot create any. Whatever protons we have in this universe are it. This will be further explained in a later section.

The Greater Universe is like a never-ending ocean of Type-0 particles. Well, the Greater Universe is never-ending but the ocean of particles at some stage, maybe after $10^{10,000}$ years or more it will get to the area that no particles exist at all. The absence of particles does not mean the universe has ended. As discussed before, the universe is not bordered and is limitless.

Figure 19: Newton's Third Law of Motion—Newton's Cradle

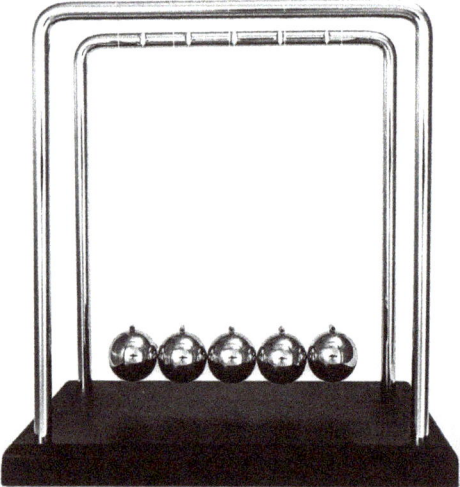

Reaction Vector

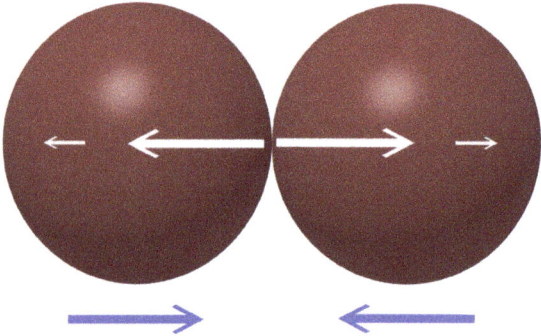

**Force Vector/
Acceleration Vector**

In this Greater Universe exist smaller, much smaller, universes like ours; about 100–200 billion of light-years in diameter. The Greater Universe looks like a Swiss cheese with holes inside. These smaller universes, not many of them, we know 100% of the existence of at least one, ours; are thousands of trillions of light-years apart. The pyramid structure is like this:

We are on our planet, the Earth, which is part of Solar System, which is part of Milky Way, which is part of Virgo Cluster, which is part of Local Galactic Group, and many of these Local Groups create the universe.

> Earth → Solar System → Solar Interstellar Neighbour-hood → Milky Way Galaxy → Local Galactic Group → Virgo Supercluster → Local Supercluster → Observable Universe (Our Universe) → Group of Universes → Universe Cluster → the Greater Universe

The Structure of the Universe

Although there is no evidence of the existence of other universes, and we will never know for sure, if they did exist, Groups and Clusters of universes are the way we would categorise them.

It is not possible to know if another universe exists. It is not possible to travel to it or receive any light or radio signals from it. The scientists have estimated the size of our universe is about 100 billion light-years across. Due to the way the structure of our universe began, which I will explain later, there is another trillion light-years of empty space after the last light we see on the edges of our universe. The scientists' estimation is based on the last light source—a star, or last planet that exists; after that, there is a very great distance of totally empty space before we reach the outskirts of the universe. If I can use a large city as an example, we get out of the CBD, drive out of town, the buildings and houses slowly disappear, the distance between houses that was, in the city, less than a metre, is now hundreds of metres. We get to a point where there are many kilometres between houses, and the next fuel is 300km away. Between those towns is nothing but desert. In a city like Khartoum, the capital of Sudan, there is nothing but sand and dust 5km out of the city.

Our universe expands across about 100 billion light-years, but the empty space is part of this universe that stretches for another trillion light-years.

Figure 20: The Greater Universe

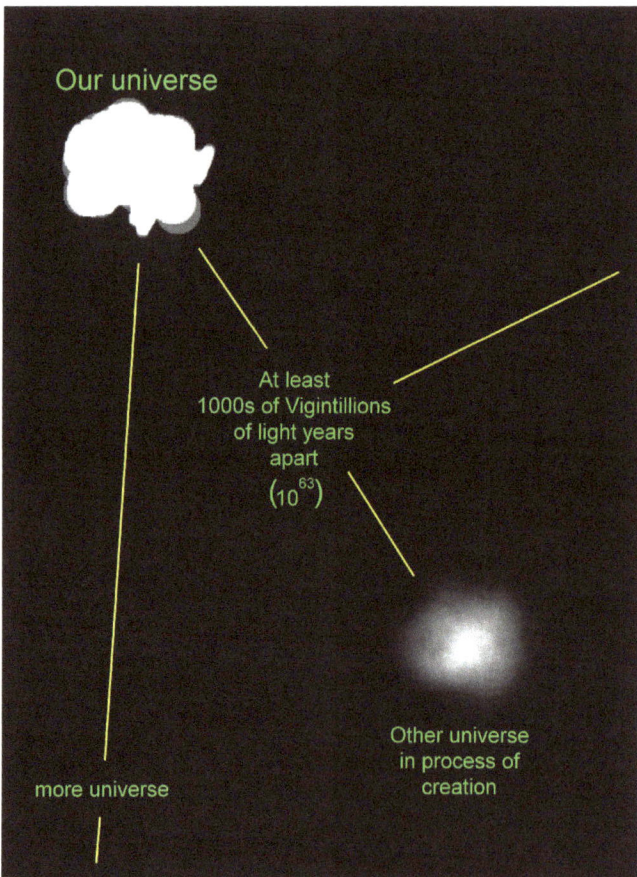

Our universe

At least
1000s of Vigintillions
of light years
apart
(10^{63})

more universe

Other universe
in process of
creation

Figure 21: The Universe Cluster

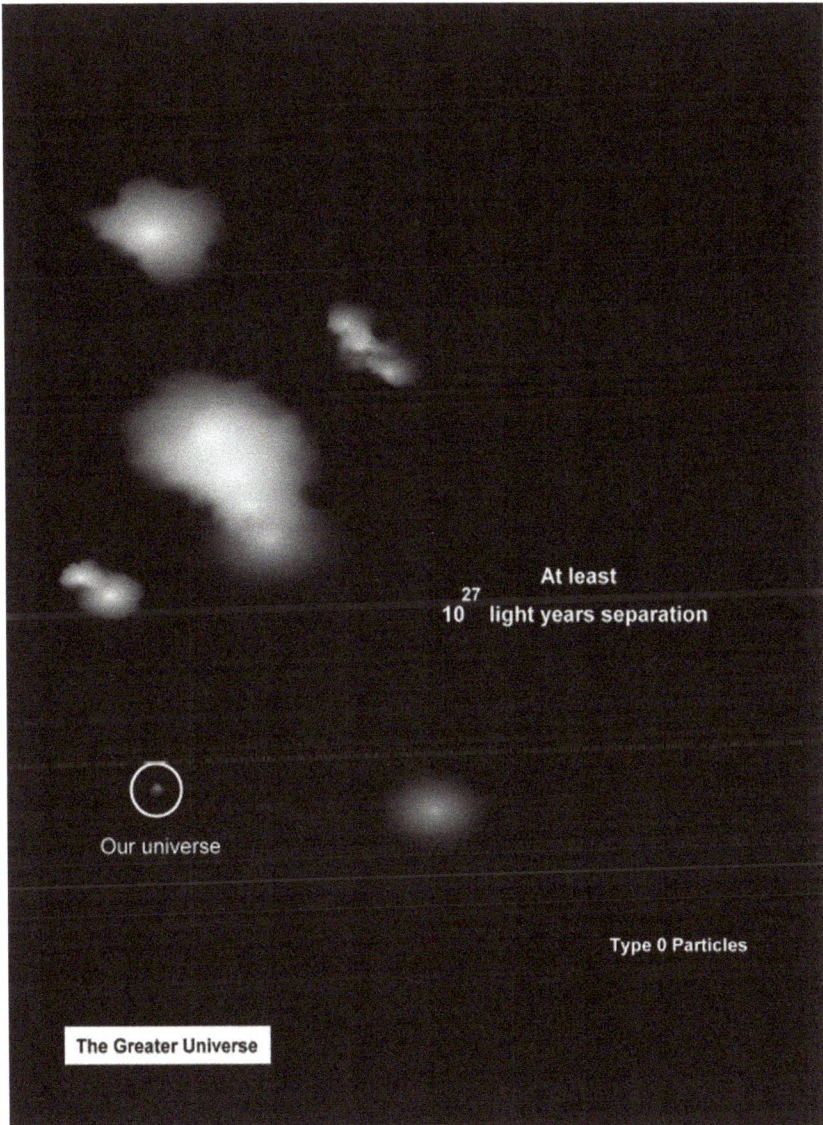

At least
10^{27} light years separation

Our universe

Type 0 Particles

The Greater Universe

Before we go further off track, let's go back to the armchair and the position we were.

Figure 22: Outside the Clouds of Particles in the Greater Universe

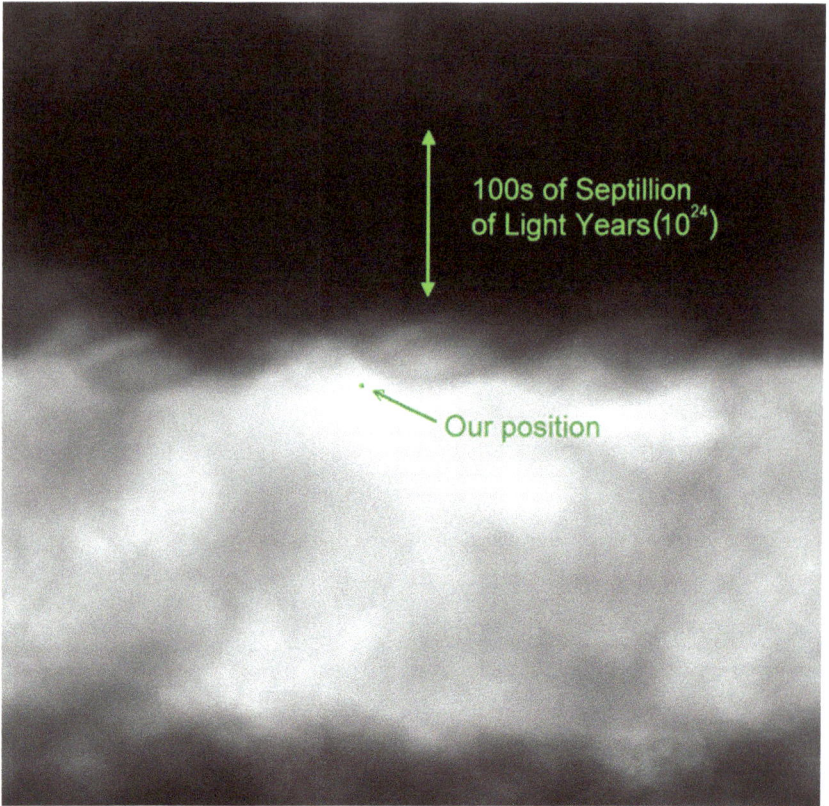

The size of this ocean of particles, unlike the size of the Greater Universe, is not limitless. We don't know for sure but, if there is more than one universe other than ours, and a cluster of universes exists, then I would put a number on the size of this ocean of particles. I would estimate the size of this would be $10^{1,000,000}$ light-years across, at least. ($10^{10} = 10,000,000,000$). This is a very large number; you might think the size of this number is ridiculous and absurd. It stretches for 2km if printed on a roll of paper. It also makes no meaningful sense. It is a one with a million zeros. We don't have a name for it, and this is the first time I've measured the size of a number in kilometres. This number is so large that printed out it would stretch from the Earth to the Moon.

My personal view is that the original size of the Greater Universe was about $(10^{1,000,000})^{1,000,000}$ light-years at the time of creation, it then contracted. This is a very large number, but it only represents the diameter of the Greater Universe.

Usually, we present the measurement of distance in numbers—not this time. The length of the number is presented in kilometres. I think some may be unhappy with the use of such large numbers, as they rightly would say they make no sense and are meaningless. If you think that is a large number, try to calculate the number of particles per cubic meter then calculate the number of Type-0 particles in that Greater Universe. The number should be equal to a return trip to the Sun. Just as an example of number of atoms: an adult is made up of around 7,000,000,000,000,000,000,000,000,000 (7 octillion) atoms, or a little more if the person has long hair.

This number gets even more meaningless if you try to convert it from light-years to kilometres (1 light year = 9,460,730,472,580 km) or calculate the volume of the cloud in cubic kilometres. We need to put a number on the size of the ocean of particles in order to understand the size of the universe. The particles don't go forever; they end somewhere.

The sheer size of the Greater Universe shows that if the particles were made by something like Eyjafjallajokull, we would have needed a lot of volcanos to produce the amount of the particles present.

Figure 23: Size of Ocean of Particles

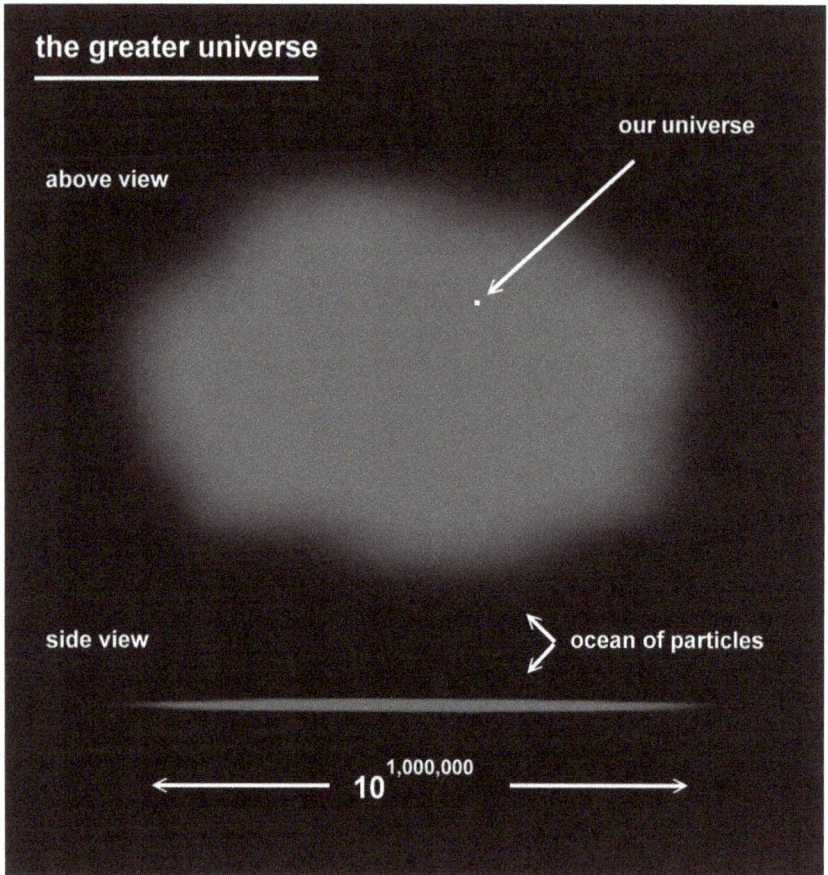

The size of the UNIVERSE is different from the size of SPACE. In my opinion, the universe is the one that includes our planet and the rest of the stuff we see when we look up in the sky. Scientists say its size is around 100b light-years. I believe our universe is part of a larger universe that stretches for $10^{1,000,000}$ light-years, at least; then all this 100b or more is located in an empty place called space that goes forever and is not limited.

Unlike Local Clusters and Universe Clusters, we cannot have a cluster of oceans of particles.

If we pull back our point of observation (about $10^{1,000,000}$ light-years) to see the whole ocean of particles in one frame, we will see, if we could because no light is emitted from it, a very unshaped cloud that has no centre core, disappears on the edges and has no particularly dense areas.

If you look at the clouds on our planet, they have many shapes, from Stratus, Altostratus, to Cirrus, or Cumulus. They are shaped this way because there is energy to push the water vapour together and create such a shape. With the clouds of particles, there is no energy to push them together and create shapes. Energy has not been created yet.

Energy

Back to our position outside of the clouds, or not exactly outside it: somewhere outside the area that will eventually turn into our universe. These particles, although the scientists call them Dark Energy, have no energy at all. Energy is yet to be created. At the time of creation of these particles, energy was not present in any form.

Everything we see in this universe is made of molecules and the subatomic particles that have energy. If the molecules are broken, and subatomic particles are separated, they release the energy that was stored in them at the time of creation. Whether it is one molecule that friction is applied to and mixes with oxygen to release its energy (a small amount of Phosphorus (P) on a small wooden stick, a match); or $C_{12}H_{23}$ (Diesel) mixed with Oxygen expands fast and rapidly releasing energy when pressure is applied—the diesel engine works in this way.

How about $C_3H_5N_3O_9$ a very unstable substance called Nitroglycerine; or the food we eat that the body turns into energy? How about the wind? The sun heats up the land and air, warmer air rises, and cold air replaces the warm air, this continues on a larger scale and wind is created, then we capture that energy by placing large fans and turn it into electricity. There are many forms of energy but they all have one thing in common, the energy is put into the substance or entities first, then it is converted or released by a different process. This process is either interference of molecules (mixing two substances) or a change in the subatomic part of the substance (creating a chain reaction such as using Uranium). When we boil water, lots of energy is put into the molecules of water to turn it into steam, and the steam can then be used to push the train or turn a turbine.

Creation of the Building Blocks of Matter

In this universe, whatever is used to generate energy, whether from creating steam or cracking an atom, we are releasing the energies that had already been put into the elements. This energy derives from the molecular structure and the subatomic formation of every element in the Periodic Table. Whether it is Hydrogen with Atomic Number of 1 (one electron) or Copernicium with Atomic Number of 112, or any other elements, energy was used at the time of creation to create the proton and the electrons and put them together. So when and where did that energy come from, and how this was done?

Everything in the universe, as far as the scientists have yet been able to discover, is made of one of the elements in the Periodic Table. We can probably say there are some more elements in the universe that we haven't yet been able to discover. Maybe there are some more with heavier atoms and higher atomic numbers in other planets across the universe; but if we do find them, they will all have again one thing in common with what we have here now. How about the Sun and other planets? In the Sun we can see there is a lot of energy but how about the planets, or the Moon or the meteor that passes the

Earth every once in a while? To some that piece of rock the size of a school bus may be a dead entity, but it is full of energy. The energy that binds the molecules, and the subatomic energy, are released when it comes in contact with the Earth's atmosphere, and the friction causes the release of that energy. Everything in this universe is made out of protons and electrons, and it took a lot of time and energy to create those electrons and protons. This still leaves the question: when and where did that energy come from; and how was this done?

Figure 24: Periodic Table of Chemical Elements

Group →	1	2	3	4	5	6	7	8	9	10	11	12	13	14	15	16	17	18
↓ Period																		
1	1 H																	2 He
2	3 Li	4 Be											5 B	6 C	7 N	8 O	9 F	10 Ne
3	11 Na	12 Mg											13 Al	14 Si	15 P	16 S	17 Cl	18 Ar
4	19 K	20 Ca	21 Sc	22 Ti	23 V	24 Cr	25 Mn	26 Fe	27 Co	28 Ni	29 Cu	30 Zn	31 Ga	32 Ge	33 As	34 Se	35 Br	36 Kr
5	37 Rb	38 Sr	39 Y	40 Zr	41 Nb	42 Mo	43 Tc	44 Ru	45 Rh	46 Pd	47 Ag	48 Cd	49 In	50 Sn	51 Sb	52 Te	53 I	54 Xe
6	55 Cs	56 Ba		72 Hf	73 Ta	74 W	75 Re	76 Os	77 Ir	78 Pt	79 Au	80 Hg	81 Tl	82 Pb	83 Bi	84 Po	85 At	86 Rn
7	87 Fr	88 Ra		104 Rf	105 Db	106 Sg	107 Bh	108 Hs	109 Mt	110 Ds	111 Rg	112 Cn	113 Uut	114 Fl	115 Uup	116 Lv	117 Uus	118 Uuo

Lanthanides	57 La	58 Ce	59 Pr	60 Nd	61 Pm	62 Sm	63 Eu	64 Gd	65 Tb	66 Dy	67 Ho	68 Er	69 Tm	70 Yb	71 Lu
Actinides	89 Ac	90 Th	91 Pa	92 U	93 Np	94 Pu	95 Am	96 Cm	97 Bk	98 Cf	99 Es	100 Fm	101 Md	102 No	103 Lr

Humans, animals, trees, rocks, the virus in the bottom of Antarctica that has been around for more than a million years, the Moon, the largest known star in the universe—VY Canis Majoris, and everything else in this universe have one thing in common: they all are made of protons and electrons, then molecules. There is a lot of energy holding the molecules together; where did that energy come from? We need to narrow the search and find one common source. We need to go to that source and explore where it comes from and how it was created. Fish, clay, apple, snake, fig tree leaf, mud, air, protein, vitamin, ... anything—have one thing in common—ATOMS, Molecules, protons and electrons.

Although this book is not concerned with the creation of fish or humans, we need to go back to the very first, very, very first, creation of the proton, or maybe the first electron, to understand the reasons behind the creation.

When were the first protons and electrons created, and how was the energy stored in them?

To answer these questions, we need to go back to our armchair position, just outside the large ocean of particles, about 100 trillion years before the Big Bang when everything began. The creation started a very long time ago. It created the Type-0 particles, now, phase two starts, and we are going to explore it step-by-step.

The Size of the Universe

A footnote before Stage 2 of creation

The common belief is that the universe was small at first, then it started expanding after the Big Bang. The universe has always been very large, long before the Big Bang. The particles in the universe first contracted to create matter before the Big Bang, then after the Big Bang, the universe started going back to its original state. The universe was never small, and the expansion we see now is not the expansion of the universe, the matter inside the universe is going back to its original shape. Only after it has gone back to its original shape and is still expanding, can we say it is expanding beyond its original size. When we say the universe was small before the Big Bang, we draw a border around the universe, limiting its shape. The universe is limitless. The universe is infinite. The universe has always been and always will be infinite.

Let's explore this idea, the universe was small as they say, like 1 billion light-years across at the beginning; what was after the 1b light-years? It now has expanded to 13.7b light-years across—into what? What would we see if we could position ourselves at 14b light-years? Is there a dead drop? The universe, like space, cannot be bordered, contained, or confined.

The definition of the universe, as a noun, can be said in a very similar manner, to the definition of a forest:

Forest: a large number or dense mass of vertical or tangled entities.

Universe: a large number or dense mass of scattered or tangled entities.

"Universe" is a name that refers to a large existence of matter and non-matter (*Type-0 particles*), including all the matter we have discovered and all we haven't discovered.

We always refer to the universe as an entity and not as a collective noun. Universe cannot be an entity because it is a title to a group of entities. The definition of Entity is *a material thing that can be seen and touched.* We cannot see or touch the universe; therefore, it either does not exist or is not an entity. We talk about the universe as if it is like an overinflated balloon with everything inside and it keeps getting bigger.

This definition of Universe is more precise:

The Universe is a collective noun of the totality of existence and is infinite.

CHAPTER 8:
STAGE TWO OF CREATION

First Laws

We are just outside the large ocean of particles in a place where the particles will eventually become the universe of which we are a part. We are not completely outside the Greater Universe, just inside, where it will become a universe in about 50–100 trillion years' time. The particles are almost in a state of plasma, very similar to the state the scientists say the universe was before the Big Bang, *plasma-like state and super-heated*. In contrast, the state of the particles is super-cold, not super-heated, and the plasma-like is just a reference to the amount of the particles. Particles can never become super-heated plasma, and the universe has never been in such state.

There is no source of heat and no source of energy; the very first energy will be created at least 40 trillion years after this point. The Type-0 particles will burn off if subjected to heat without leaving any residues or release of energy. If the temperature rises from −8,000°C, as it is now, the particles will just disappear. These particles are not an element of any sort yet, not even an element we haven't discovered. The closest thing I can describe is they are like the very small snowflakes we see in blizzard conditions. The snowflakes are not touching each other despite the close distance between them, and there are so many you can't see more than a metre; and when the temperature rises the snow disappears.

Charcoal balls or cubes are made from charcoal powder. The charcoal made from trees is turned to very small particles then pressed by a machine and turned into cubes. The charcoal powder or particles burn very quickly with high intensity if exposed to heat; the reason is that all areas of the particles are exposed to heat and oxygen at the same time. On the other hand, if they are compacted into a cube only a very small part of the particle is exposed to heat and the heat is dispersed through the cube, so a lesser temperature is applied to the particles and it slows the process. The cube burns very slowly and produces less heat. A cube can burn slowly for hours, but if made into particles all of it can burn in seconds and release a lot more intense heat.

Figure 25: Charcoal Cubes

Type-0 particles in space are the same as the charcoal particles. They burn (burn is not a correct description; particles just disappear if the temperature is increased) very quickly—I should emphasise they don't burn in a similar manner to burning on Earth, there is no Oxygen or any other element to allow the burning process.

I don't see how the ocean of particles became super-heated in a plasma-like state and exploded as the scientists say. I cannot see the state these particles were in could have been as the scientists say they were. Any increase in temperature at this stage (50–100 trillion years before the Big Bang), would just cause the Type-0 particles to disappear, they don't even get to the point which they would become super-heated. They just cannot sustain temperature increases unless they are compacted—similar to charcoal cubes, into larger particles like protons. The creation of the first proton is 40–50 trillion years away.

To understand the process of creation we need to follow the laws of physics. We need to go backwards from the Big Bang and see when and under what conditions those laws were created. Super-heated, plasma-like; when was the heat created, how did it become plasma, and what was the process? How long before the Big Bang was the heat was created, by what, and what was the source of heat? These are what I would like to know:

- Heat
- Source of heat
- Energy
- Plasma state
- Time
- The explosion
- Movement

What triggered the explosion that resulted in a chain of events and the Big Bang? It is not as if someone was doing some maintenance using a grinder, or welding something, and accidentally blew up the place. It certainly wasn't a star or planet that travelled for some time and crashed into the super-heated plasma; and, definitely, there was no magic wand.

Movement or motion is time if it is constant. How long did it take the particles to move closer to each other and create the plasma? When were the first laws of physics created and under what conditions? Where did the first energy come from which moved the particles closer and where was the source of heat? When and how was the first energy created?

If we want to know how the creation started, we need to know how the laws of physics started. Physics is the answer to the creation. By finding the first laws of physics, we can find the first creations.

Motion

Motion is the second law of physics created in this universe. Something was moved from A to B and must have obeyed the law of physics' or Newton's Law of Motion. It needs a source, and the source needs energy. One can argue that the laws of energy were the first laws. One may say the creation of Type-0 particles was the first law and the conditions in which they were created. These conditions can be anything, and they require some energy. Alternatively, the first law of physics was Mass, and it was created in the first phase of creation. Type-0 particles have mass, and this was created trillions of years before, when the first particles were created, right at the very beginning of Stage 1: the creation of Type-0 particles.

The first six chapters will explain how the particles were created without any of the laws of physics. Even if any laws were created, those chapters would explain how, and those laws that were outside the universe are being created now. In here, I'm trying to detail the creation from phase two. We are talking about a lot of particles that were here for quite some time. For trillions of years, since the creation of the particles, they existed and just rested there. They were never moved or disturbed in any way.

Condensation has just begun. It is the reason behind the movement of particles. The particles have started to move closer to the parts of the ocean that are colder than other places. There has been a drop in temperature. It is an extremely rare event for the temperature to drop lower than it was earlier. This is the first event in phase two of creation, and it starts the creation of universes, our universe in this case. If we believe other universes exist, not in the same form as ours, then this is what started them. This is the first step in the creation of our universe.

I cannot stress this matter enough; the only fundamental and first event that led to the start of our universe is that the particles moved closer to each other, maybe by less than one micron, but a move is a move. This first step is, in fact, the one required to start any universe. The creation of Type-0 particles was the first step in the creation of the Greater Universe, in this universe the movement of particles was the first. The details of the creation of Type-0 Particles and the Greater Universe are not discussed here. The creation of the Type-0 Particles was the second event; the first was the creation of the conditions that caused the creation of the particles.

We are here to witness the creation of our universe.

This is the second time temperature has dropped. This will also be the last time. The temperature will never again drop to this level; this cold snap will last 20–40 trillion years. The temperature hasn't dropped in the entire ocean of particles but only in some areas. The drop in temperature started very recently, about one billion years ago. Although these patches are thousands of light-years apart, it has come at almost the same time across the ocean. The distance may seem a lot to us, but in the Greater Universe, it is not important.

The rate of movement is so small it cannot be measured. It is maybe a millimetre every 100 years, but it is a movement. This is the first time the particles moved. The distance between the particles can be measured in millimetres. At some points, it is almost zero, but they are not touching; there are gaps in between. The particles are not in any particular shape and are forming a shape like a liquid sea.

The particles are not pulled by any forces. In situations like this, at the current time, gravity is the source of the energy to move a particle to the centre of the event. At this time of the creation, about 50+ trillion years ago, gravity is yet to be created, and no other form of energy has been created either. The creation of the first energy is at least 40 trillion years away. Also, gravity works only to a certain distance; it is not effective all the way to the limits of the new universe being created. For example, our Sun's gravitational forces are limited to within our solar system, once you pass Pluto the force disappears. The gravitational force also needs a mass; the larger the mass is, the stronger the force is. The mass is made of molecules and atoms like a large planet; the creation of the first ++atoms and molecules is at least 40 trillion years away.

Instead, the particles are directed towards the centre of the event. The directing or pushing requires energy. It is not energy that is bringing the particles together; it is the low temperature. The move is not like a black hole that sucks everything to its centre, if it were, all particles would have been directed towards the centre of the new universe that was being created. Instead, the move is more localised, and the particles are moving smaller distances towards each other rather than long distance to the event's centre.

This part of the Greater Universe is about to start contracting. In order to create a new universe, this part of the universe needs to contract or come together. The size of this part, which will eventually become our universe, is billions of times larger than the created universe we have now. It is billions of trillions of light-years across. The contraction of this will leave a very large distance of empty space at the end of our universe to the rest of the Greater Universe. If we could travel to the last part of this universe, to the very edge of this place and look back, we would see many dim lights emitted from stars far away and trillions of light-years of absolutely nothing ahead.

This contraction doesn't mean the particles from other parts of the Greater Universe will move and fill the void. The particles are not attached or attracted to each other, and there is no energy between them. If a particle or large amounts of particles move from their place, that place will remain void and not filled. It is not like a pool of water you put your hand in and move it around and the water behind you hand will fill the empty space which is created by the water you pushed away.

First Law of Geometry: Angle

The motion brings the particles closer, and sometimes they start touching. In the first instance when three particles come together they create the first law in geometry, they create an Angle.

Some would argue Length was the first law of geometry like the length of the particles; they might be correct if we talk about the creation of the first particles and the rest in the first six chapters of the creation. In this universe, ours, the angle is the first law of geometry. This angle though can't be seen, but it happens as particles physically touch each other at the same time, and for a very short time. They move along and separate again, but the first angle is formed and could be measured in degrees with a protractor if we had one.

Some of the laws of physics may have been created before this time and at the time of the creation of the particles. It is not clear yet, but, in the process of creation of this universe, trillions of years after the creation of the particles, laws of physics are further created for the first time. This is what this book is all about, it will explain some of the first laws of physics created in this universe, and after you read the first six chapters you will see the first laws in this universe are also the first laws of the Greater Universe too.

We have always looked at the sky, and drawn lines between the visible stars and named them. We drew lines between the Crux stars:

Gacrux, Acrux, Becrux, Delta Cru, and Epsilon Cru, and called them the Southern Cross. We name them and identify them as we draw a line in the air between them; from our point of view, they create a cross that is visible from the southern hemisphere. These stars are at least hundreds of light-years apart, and from a different location, they do not create a cross. It is we, the humans, who look at things, draw lines, and identify shapes.

A little later, the second law of geometry is created in this universe, Length. It happens after more than two particles line up and create a straight line of particles that is continuous.

Figure 26: Southern Cross and Night Sky

Figure 27: Human Interpretation of Night Sky

At this time of the creation, these particles had never moved before and certainly never came close enough to touch each other. The earlier motion of particles brought them together and created a physical shape that we call an angle. Although it is a very small angle, in fact, the smallest angle ever, it can be measured. If we had the tools, we could measure the angle in degrees and say, for example, it is a 72-degree angle.

Space as a Vacuum

The space at this point is at zero-pressure; therefore, there is no energy whatsoever to create any form of pressure on the entity. It is important to point out that space has always been at zero-pressure. Some may refer to the area of outside the Earth's atmosphere as a vacuum, or count anywhere outside the space vehicle as a vacuum. My understanding is, we live in a pressurised environment, and anything outside should be considered as zero-pressure. The pressure of this environment comes from the weight of the atmosphere and the weight of the molecules in the air, and this weight is caused by the earth's gravity.

The definition of Vacuum is *a space entirely devoid of matter.* As we know, outside of the earth, space is not entirely devoid of matter.

In scientific articles and literature, a perfect vacuum is referred to as a 0Pa (Pascal) pressure. The definition of Perfect Vacuum is *an ideal state of no particle at all.* What if we create a chamber that holds an absolute vacuum at 0Pa and inside the chamber is a fistful of sand; would it be considered as a perfect vacuum? No, it wouldn't, because, as the definition says, *state of no particles at all* and the presence of sand grains will override the definition although the pressure is at 0Pa.

Only at the sub-atomic level, does an absolute vacuum exist. The space between the proton and electrons is void of matter, and no other particle can enter this area unless it is a Type-0 particle. There is also no pressure applied.

In space travel, the area outside the space vehicle is commonly considered to be a vacuum. What about the matter that is pushed into space by the rocket's engines? The space rocket leaves a trail of matter or the burnt fuel behind; would it still be considered as a vacuum by definition? The only time a state of vacuum, by definition, was present, was at the time before the creation of Type-0 particles, which goes back to trillions of trillions of years ago. Since the creation of particles, trillions of years ago, the universe has not been entirely void of matter therefore not a Perfect Vacuum.

Still, to this date, there are pockets of the Greater Universe in excess of trillions of cubic light-years in which Type-0 particles and pressure do not exist and that are a complete void or a perfect vacuum. This area, the trillions of cubic light-years, is only a small part of the Greater Universe. It is like a small bubble of air floated in the Pacific Ocean, because of the bubble we don't consider the ocean void of water and empty. The comparison of these empty pockets in the Greater Universe is similar to the pocket of air found in the Pacific Ocean. The Greater Universe is so large the void pockets represent a tiny bubble.

On earth, a vacuum need to have a source similar to a pump, and if outer space was a vacuum then where is the source, and why is not everything sucked into the source? The fact that the atmosphere is not sucked out shows there is no presence of negative pressure, there is only one force applied to our atmosphere, and that is its weight, generated by the earth's gravity. A vacuum environment needs to be contained. Where are the containment walls? We cannot call an environment that has a different pressure value a vacuum. Passenger jets pressurise the cabin for the comfort of the passenger: this doesn't mean outside is a vacuum. Low pressure—yes, not a vacuum. This low pressure will eventually reach to almost zero once the altitude reaches around 1000km. The last layer surrounding the earth is the Troposphere, and in this layer, the atmospheric activities can reach up to 800km when the Sun activities are high.

Although we might experience and register large pressurised areas in space similar to our atmosphere's highs and lows, it doesn't mean space as a whole is pressurised or is a vacuum. On Earth, the standard atmospheric pressure at the sea level is 1,013.25hPa or 14.695psi. For every 300 meters of altitude, the atmospheric pressure drops by 4%, and it reaches zero once outside the atmosphere, but the pressure value does not go negative. If space were a vacuum, it would have sucked the atmosphere out of the earth and any other planets and stars.

A vacuum should not be considered in the absence of molecules or pressure. The term Vacuum should not be used in describing space's environment. It is used very loosely and is a grey definition, which should not be used when it comes to space. It should be called a zero-pressure or neutral-pressure environment.

The concept of a perfect vacuum is something that should be looked into when the first six chapters are compiled. At the time before the creation of Type-0 particles, when the whole Greater Universe was void of matter, and 0Pa pressure, it would have been the absolute vacuum the dictionary defines. If there was a source that created the particles, then it also cannot be considered a perfect vacuum as the presence of the source will not meet the definition.

A perfect vacuum cannot be created in a lab. There is always some matter left in the chamber that cannot be completely removed. A small number of molecules of air will always be trapped in the empty space between the molecules of the walls of the chamber, and will never be removed. This will prevent it being a perfect vacuum by definition. Perhaps by use of an electromagnetic pulse, the matter can be washed out of the chamber, but some will still be trapped between the chamber's molecules. Scientists believe the Dark Matter exists everywhere in the universe and it is the fundamental part of what created the universe. Type-0 particles are not Dark Matter because they are not Matter. Type-0 particles had existed before matter was created, although matter is created from Type-0 Particles, they cannot be classified as matter as they don't have properties. Matter can and does have properties, not Type-0 particles. Dark Matter is larger in size than Type-0 particles. Dark Matter is, in many ways, similar to Type-0 particles but is in a different class. If we were going to classify the particles in size, it would be like:

> Nothing can ever exist → Type-0 Particles → quarks → Dark Matter → Isotopes → electrons → protons → Neutrons → Nucleus → Atoms → Molecules → the rest...

The group of Dark Matter, Isotopes, electrons, protons, and Nucleus are in Type-1 Particles category. Nothing after this group is considered as particles. It can be said we have two types of particles: Type-0 and Type-1. Type-0 particles and quarks are very close in size but different in properties. Type-0 particles have no properties.

The "creation of protons" section explains how Dark Matter was created.

Later I'll explain how the Type-0 particles became protons and created the universe we know. The Type-0 Particles don't exist in this universe anymore as freely as believed. A very small amount, leftovers from the past, exist and perhaps one day we can get hold of some; but they don't exist everywhere outside the earth and in space around. If the Type-0 particles exist in a large volume, as is believed, then we could not have a crystal-clear view for the distance of 13b light-years. As we know the Type-0 particles are smaller than light particles and light bounces off the surface of the particles and cannot pass through them, so, if there are a lot of particles around, we are very likely to see very dark clouds with no light whatsoever coming from them. If there are large fields of particles floating in the universe, the likelihood of being able to see them through a telescope is very high.

There is a very small number of particles left in our universe after the Big Bang but not anywhere near our solar system. I'm sure if we could mine the Moon to its core we would find some. Sometimes the asteroids may contain some at their core remaining from when they passed through a cloud of particles billions of years ago and collected a small amount. The particles, after millions of years, find their way to the core of these entities as the gravity slowly pulls them down.

The Particles are Moving

We continue our trip on board our armchair; we push forward for thousands of years. Slowly, we notice particles accumulating and pockets of empty areas. These areas are metres from each other. The void pockets are created by the moving particles. These particles have moved to different places and have created areas that contain large numbers. Passing through the area is not slowing us down, as there is no resistance from the particles. Resistance requires energy and the energy in any form has not been created yet. The rate of movement of particles is still very slow and will not increase for the next several million years. The empty pockets continue to grow. It is very difficult to see the accumulating particles; they are like a large amount of debris on water with no particular pattern.

Figure 28: Start of Particles Formation

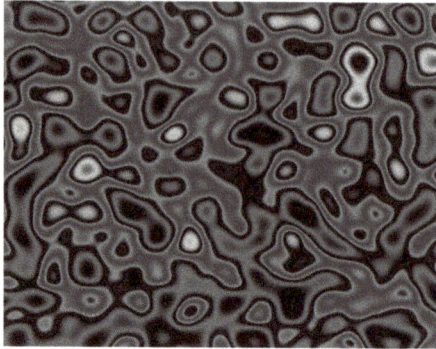

We push forward and travel another million years. Greater accumulations of particles appear, and the empty pockets are becoming larger. The particles are now touching in accumulated areas and the distance travelled in those areas takes hours, not days. This condition continues to progress for another billion years. At this time, we see clear areas with no particles at all and areas congested with particles. It takes at least a day to travel through an area. These areas of particles are still brought together by the condensation. The area outside the newly created universe is slowly becoming visible. We can now draw a line between the old Greater Universe and the part which is contracting and creating the new universe, ours.

If we look at the Greater Universe from far off, we can see very fine line-like cracks appear as in a dried land. We need to be trillions of light-years away to be able to observe this event. These cracks on the outer edge of our universe eventually widen to at least a trillion light-years. This gap widens at a very slow rate, and it will not become a completely void area. Particles slowly disappear to almost nil, but the area will not be an absolute void.

Figure 29: Line-like Cracks Appearing

The particles are still moving at a very slow rate and therefore the shape of the area changes very slowly. The empty pockets sometimes remain empty for hundreds of thousands of years then get smaller, larger or just change their shape as the particles move around. Empty areas fill with particles, and new empty areas are created nearby. The speed of movement has not increased and will remain for another trillion years. It seems the creation is in no rush and will continue moving in first gear for a long time.

Our Universe is Created

We are now past a billion years and the gap, between the Greater Universe, and the newly created universe, is wider. Before, it was a one-piece universe; now it is clear a new universe is born and will become our universe, as we know it now. In the past billion years, the particles continued moving towards each other and created a rather more populated area. The gap between the two universes is narrow, but it will continue widening as the particles continue moving away from the rest of the Greater Universe.

Inside the new universe, the particles are still moving but not towards the centre; there is no pulling force to drag them anywhere. There is no force or energy yet.

At this time, the size of the newly created universe is trillions of times larger than the universe we have today. The universe is contracting, and the speed is still at the lowest rate. This process will continue for another 100 billion years. Now we leave the Greater Universe and concentrate on the new universe in the process of creation. Nothing of any immediate interest will happen to the rest of the Greater Universe anymore. The gap is the containment line that separates the new universe from the old. Nobody will ever know if any other universe has been or will be created from this point, but if it will, or has, it will be created by a similar process. The Greater Universe will remain exactly the way it was and will never change. Imagine a very large amount of particles will float in the space for the rest of the eternity.

Sometime after this point, the particles start to move a little bit faster towards each other. The increase in speed is so small it cannot be measured by any instrument. At least a billion years from this time the increase in speed constitutes ACCELERATION. The second law of physics is created.

Second Law of Physics: Acceleration

As Newton says, the first law of physics is the Law of Motion. Acceleration is second. When we arrived at Stage 2 of Creation, the first law of physics had already been created. The motion of particles was the only thing that happened in the first stage beside the creation of the particles. There was a very long pause, then, we arrived at this stage. The law of motion was carried over to this stage, but

Acceleration is the first law created in this stage. The motion of particles in the first stage was very small, and it did not apply to all the particles. A very small number of particles moved and this was not significant in total and stopped after a short time. In stage 2, we see the motion is more widespread and continues. As the result of this motion, we have the creation of angle and length.

Mass, another law of physics, was created in the first stage and it was the mass of particles although very small for each particle and mass of all the particles in total.

Acceleration occurs when an entity at zero-point motion (stand still) starts moving, the relative motion is changed, and the change is positive. If the change of relative motion is negative, it is called Deceleration.

The rate of a particle's motion is so small it cannot be detected by any device, and, if it were observed, it would take millions of years to observe the movement, but a movement is a movement. Isaac Newton didn't say it has to be at least some specified units of measurement per unit of time. If the particle has vector quantities (magnitude and direction), it qualifies as moving.

Third Law of Physics: Velocity

This acceleration continues for billions of years before the particles gather speed. The result of mass and speed is Velocity.

Another law of physics is Velocity, and it has just has been created for the first time. Velocity is magnitude and direction between the initial and final position of an entity. The particles have speed, and they're all moving in a direction. The initial speed and the final speed of the particles, although small, differ, and it indicates the velocity by definition. The particles have mass, and the rate of (velocity) speed and direction is not changing. If the velocity is 10m per second, it will remain at 10m/s and will not go to 11m/s then drop to 9m/s. The incremental speed remains the same, so the velocity speed also will remain unchanged. This velocity is a Constant Velocity, and the rate of Relative Velocity is zero. Constant Velocity means the entity maintains the speed and direction at the same rate. Relative Velocity is the difference in speed and direction to the other entities in the same group. All particles have started to move at the same time and rate in the same direction. All particles have the same vector.

It will be billions of years for the velocity to be high enough to measure, and, by then, the particles start colliding with other particles moving in a different direction. This collision includes both types of particles, and they collide head-on with other particles and non-head-on.

The head-on collision has a clear definition, evident from its name. It is referred to when the velocity of two entities are along the path of impact. The second type of collision, or non-head-on, occurs when the particles had a head-on collision and then changed direction as a result and collide with other particles. The velocity of the two entities is not along the same path of impact.

Density

The collisions continue for billions of years and create Density. After the collision, the particles stop and stay where they are. However, the moving particles from behind keep coming and add to the number of the particles in this area. Density is a relationship of mass and volume (Density = Mass/Volume ($\rho = \frac{M}{V}$)). For example, density of 1 unit of mass in 10 units of volume is 0.1; and the density of 10 units of mass in 1 unit of volume is 10.

The area, which was occupied by fewer particles, is now occupied by a large number. The particles at the time of creation had about the same distance between them. There was a small amount per cubic unit of measurement. Now, trillions of years later, it has become denser. The density continues increasing for trillions of years. The particles continue to move from other parts of the cloud to the area that has become the centre of their colony. Like water splashed on an oily surface, the particles start creating broken areas as opposed to a continuous pool of particles. The distances between the colonies become obvious and continue to grow. It will be at least another trillion years before anything happens.

Figure 30: Greater Density of Particles

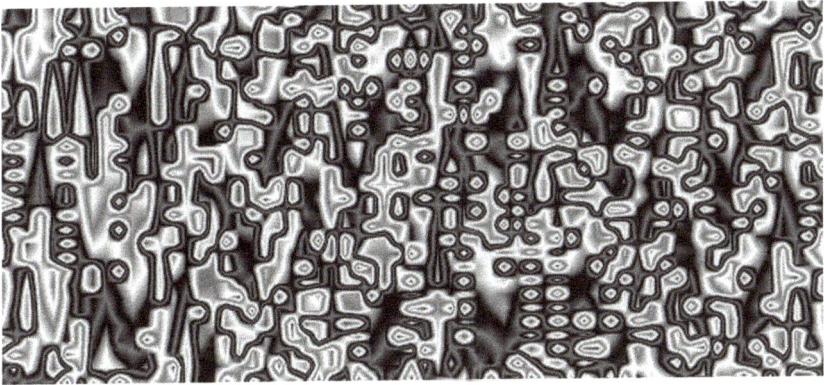

The colonies still have no specific shape or size. The particles continue to move to the centre of their local area in no particular order. At their slow rate, it will take trillions of years more before the colonies start joining other colonies, creating larger colonies. The distances between the colonies grow, and now a clear path between the colonies appears.

In the next trillion years, the paths and colonies collapse when the particles move closer to create even larger colonies. The larger colonies grow and become even larger. By this time, the size of the largest colony is smaller than our moon. The migration of the particles continues, and the speed of the moving particles increases notably. If observed from a distance, you could see a moving black cloud.

This migration continues for another 10 trillion years. The mass of particles becomes very large, by this time each of the colonies, and there are quadrillions of quadrillions of these colonies, are larger than our Earth.

About 12 trillion years later density started from the accumulation of particles in one place. For the first time in this universe, Gravity was created.

Gravity

Gravity, or gravitational force, is the tendency for all bodies of matter to attract and pull each other. Gravity became part of protons and electrons at the time of creation. The amount of pulling is so small it cannot be detected by any instrument unless the protons and electrons create atoms and molecules and it can be measured on a large scale. The amount of gravity is directly related to the density and mass of matter. The larger and denser the matter is, the greater the gravity. Gravity comes from protons and electrons, not from the formation of molecules. Black holes have extreme gravity due to the density of the mass they are made of, and the mass is denser at the core of the black hole. The planets in our solar system are in orbit due to the gravitational force of the Sun.

With gravitational force comes Potential Energy, or gravitational energy. Potential energy exists only within a gravitational field. It means that, if you move an entity in the opposite direction to a gravitational force, you are storing the same amount of energy in the entity, so the entity can use it to move back to its original place. If you remove the gravitational field, the potential energy in the entity is removed. For example, if you lift a book from the floor and place it on a desk you have stored enough energy in the book to allow it to move back to its original status/level of its own accord when the table is removed. If you move the book to 5,000km above Earth, the potential energy is still present in the entity, but if you remove the gravitational field, the potential energy disappears (removing the gravitational energy requires the removing the Earth). If you move the entity further, it will be outside the gravitational field of the Earth, and the potential energy is removed.

We need two descriptions for the same energy because, in larger areas of the universe, a star with a very powerful gravitational energy stores potential energy in another planet also with powerful gravitational energy. The planet's gravitational energy increases by the potential energy, and it starts pulling other planets, or other entities towards it. Until now, it couldn't pull them as it had lower gravitational force. Potential energy is gravitational energy but from the subject's point of view. It is the same amount and same form of energy. One is discussed from the owner of such energy and the other from the recipient of the same energy. The amount of gravitational energy of the recipient of potential energy is increased by the amount of the potential energy.

Gravitational energy only becomes potential energy if the recipient subject is moved in the opposite vector to the source of gravitational force. For example, on Earth, everything is subjected to gravitational force from the Earth, the Moon, and the Sun. At night, entities should be heavier in weight than in the daytime. This is due to the combination of Earth's, and Sun's gravity, and in daytime, the same entity should be lighter because the effect of Sun's gravity is pulling against the Earth's gravity. This difference causes tides. As you are aware, the tides are a direct result of the Moon's, and the Sun's gravitational force.

At this time of stage 2 of creation, the particles have mass and density enough to create gravity for the first time but the amount of the gravitational force is not enough to pull a lot of particles closer. It will be billions of years for the mass and density to increase and gravity to rise. When gravity increases, it affects the density, and it creates denser colonies. The density continues to increase, and it helps the colonies to look more like an entity than a compressed cloud of particles. Although there are still a lot of clouds of particles present, now, separate entities appear. These entities have nothing to do with the real entities we have after the Big Bang; they have no properties, and they are not, really, one entity: they are more like a dandelion flower, which is easily blown apart. Density continues increasing, and the entities now have a highly dense core, so will not be blown away.

The entities continue to get bigger as the gravity helps to increase the size of them. The entities are now large enough to be observed by the naked eye, and, for the first time, we have a diameter, and the entities can be measured. The size of the entities will grow in the next few billion years. The large entities, still in no particular shape, start colliding as the gravity increases. The bumping into each other creates force.

Force is a pull and push interaction between entities and it can cause a change in the velocity of the entities and their vector. A force has magnitude and direction, although the amount of the force is very small and it cannot be measured, it is enough to change the vector of an entity. The entities also are so small and have almost non-existent masses, so need only a little force to be pushed. The amount of force is not in debate here, it is the creation of force, and it meets the definition.

The Oxford dictionary defines Force as *an influence tending to change the motion of a body or produce motion or stress in a stationary body.*

The magnitude of such an influence is often calculated by multiplying the mass of the body and its acceleration. Force will play a critical role in the creation of protons and the rest of the universe. Everything started with a little movement, which became the first law of physics.

Force will take it further from here. Force is the beginning of the creation of energy. Without force, we cannot have energy. At this stage of creation, force is the mother of all energies, one way or another, directly or indirectly.

Force and gravity work together to create even larger entities. The entities that were created earlier sometimes collapsed due to the force applied, and dispersed. Their particles could later join a larger entity and create an even larger one.

The creation and destruction of these entities continue for billions of years, until, due to the large size of entities, large collisions occur. The large entities are now the size of a small planet, soft outside and dense in the centre. The core of these entities is getting much denser every time they collide.

The collisions sometimes create a larger entity and sometimes cause it to break up. Mostly the core of the entity remains intact and joins other entities. Slowly the new dense core of the old entity that had just joined the larger entity makes its way to the core of the entity and joins the original core, so it creates a larger core. Sometimes the newly joined core stays on top and becomes the new core of the larger entity as other entities quickly join.

Over time, the core of the large entity becomes larger and larger, and the whole entity becomes compressed and dense, similar to a planet. Still, it keeps attracting particles, and the size of the entity continues to grow. These entities, trillions and trillions of them, are now the size of a planet like the Earth. Suddenly a collision between two or three of these entities creates Friction.

Friction

If one law of physics stands out as the mother of energy, it would be Friction. Sure, we cannot have friction without motion, and *friction is the force resisting the relative motion of solid surfaces, fluid layers, and material elements sliding against each other.* Without motion, we cannot have friction; but it is friction that creates energy. We can have motion, and it does nothing if it keeps moving without disruption, but friction converts the force of the motion into energy when resistance is applied. In space, you can fire a gun and, theoretically, the bullet can travel forever, as there is no resisting force to stop it. The initial force continues to push the bullet forward if it is not interrupted by any external interference.

Friction is a game changer. If it were not for friction, creation would not have continued. If we could sum up the creation into one word representing the answer to what started creation, it would be "friction". Friction is the jump-start of creation; only friction and nothing else could have started creation. The unique elementary and primary event we can zero in on when it comes to the laws of physics and creation is this point. Nothing else matters.

In the anatomy of creation, this is point zero. Without friction, the creation cannot go forward and cannot begin. The answer to the question that the first human asked, millions of years ago, of how this was created, is friction. The readers may disagree with some of the things said here (although you will, eventually, agree), but they cannot disagree with the fact that energy must have a source, and friction is the source, so it makes it the mother of all energies.

The answer to "Why and how did Creation begin?" is **Creation was an accident caused by Friction.** Creation of Life, our existence, and the rest, are the result of this accident. This is an ongoing event, and future creations will occur. There was no explosion. The explosion is the result of a sudden release of energy. Energy had not been created and, therefore, did not exist, at this stage. This is the genesis of creation and the reason for existence. This is still the zero-point energy. Despite the friction, still, no energy is created. Collisions will become more often which will lead to more powerful and stronger friction and eventually release energy.

Later, with the progress of creation, there will be new energies with new sources, but the very first energy had friction as its source. You may disagree with the timetable here and come up with your own version; at some stage, you must state the point zero of the start of the energy, you will arrive at the same conclusion. We must zero in and find the very first source of energy.

This is a turning point, at this time, and the creation begins to accelerate. From this moment, the events start taking place at shorter intervals and soon we have energy created.

The collision between the two planet-sized entities takes a long time—almost a year—to complete. The entities are pulled towards each other by gravity, but the speed they pick up is very low due to the small amount of gravitational force. The entities start colliding in a very slow motion sequence. The outer areas of the entities are not hard and dense, they look like a cloud rather than a solid body, and there is nothing to stop them. Large masses, not so dense on the outside, move into each other. The core of the entities, although small, miss the collision due to the last change to the vector of the motion, something pushes them away from each other, and the cores fail to collide into each other head-on.

At this time of the creation, the movement of the large entities is very clear. If you stood far away from the entities where the event was taking place, you could see the movements are more sequential and predictable. You can see where the entities are heading, and what will happen next, or which entities are going to collide. It is like watching a game of billiards in 3-D at a very slow speed. Looking at the recurrent events, you can calculate time and "Time" has started. The start of time is very short-lived, and it ends after a few thousand years. It almost didn't happen. One of these entities started to travel on the same path for a while, similar to the orbiting of a planet. Of course, it wasn't orbiting anything, but due to the motion vector that was created by collisions, it continued travelling on this path. It wasn't long before it collided with yet another entity and it changed its course, so the time also ended.

The collisions continue to happen for millions of years. Entities continue to collide without the cores colliding. The collisions are not without results; the entities integrate and continue to form larger entities. Finally, one of the collisions leads to a head-on collision between the two cores. The cores are larger in mass and denser than ever before. This time the collision creates friction for a longer time. We had friction between two or more cores before but it was very short, and less force was applied.

This friction, at this time, creates heat, and heat is energy and has a source. But it is so short and again fails to produce any results. The heat created by the friction is a dismal amount and disperses quickly due to the extremely cold temperature of the universe. It is just the beginning of such activities. The amount of the energy also is so small and cannot be considered as the creation of energy. If it needs to be larger, the collisions need to be more frequent and more direct.

These collisions continue and the masses of the entities get larger and larger. With larger mass, we have a larger amount of gravity. The size of the entities is not critical to the gravitation force as the entities are dense enough to create a stronger gravity. The entities still look like clouds and slowly fade away as they stretch out. Their motion is affected by the gravity and becomes more direct. We can see more large entities in sequential motion at times, and we can calculate the time base of their movements; it is a measurable period. Time starts and ends many times while the entities continue to move and collide. Space has become a very busy place, too many entities, and too many collisions; it is still dark and cold. This continues for billions of years.

The interval of the starting and finishing of time becomes longer. We no longer have time for a period of thousands of years; we have tens and hundreds of thousands of years. The entities orbit the same path, which enables us to calculate time. Although time will end soon, by definition time has been created.

Time and Space

Time is only a unit of measurement that allows us to mark periods and events. Possibly human's number one obsession is with time. It appears that all humans have a very mild OCD (obsessive-compulsive disorder) with time. We constantly measure it by looking at watches, cell phones and many other means. Before cell phones, people looked at the time just a few times a day, but now, with the cell-phone we all carry, we check the time every few minutes all day. We measure the time for the start and end of an event, but in recent times, it has become a non-occurring event. We just look at a time measuring device for no reason, and if we don't have the device, we try to guess it by looking at signs of time such as daylight. This obsession is fuelled by other events we take an interest in, such as sporting events and more so by movie makers.

We are constantly told we can travel back in time. Sometimes this is even supported by scientists too. The false promise of time travel always gives us hope we can someday go back and correct the mistakes of the past, but the next day we repeat the same mistakes. We look back so much it stops us from looking forward into the future where we constantly fail. We are always interested to know where we came from rather than where we are going! We sacrifice the future for the past. We should forget the past and focus on the future.

Usage of time is exploited by movie studios and fiction writers. Time is usually considered as a block of events, and we allow ourselves to assume we can go back to the beginning of such an event to observe or make changes to the event. Maybe if we have a big telescope and place it one day away from the earth we can see what happened yesterday; apart from this, we cannot go back in time. Time travel is a good plot for movie makers but not good news for travellers. One day's distance from Earth is four times the distance from the Sun to Pluto. It takes six hours for the Sun's rays to reach Pluto.

In Einstein's Theory of Relativity, measurements of various quantities are relative to the rate of change of observers, in particular, Space contracts and Time dilates.

Time dilation is a difference in elapsed time between two events as measured by observers either moving relative to each other or differently situated from gravitational masses.

Space Contraction is the phenomenon of a decrease in length measured by the observer, of an entity, which is travelling at any non-zero velocity relative to the observer.

Space and Time should be considered together and in relation to each other.

The speed of light is constant to all observers.

The Theory of Relativity is a law applicable to length in space and time in relation to different observers. This leads to Spacetime (Space-time, Space Time) when an observer travels in a straight line with Time and moves towards Space, this makes the time shorter for the same rate of speed (speed of light 299,792.458Km/s). This move from Time to Space can potentially reduce the Time to zero when the traveller completes and moves to Space; it also is possible to go beyond zero towards the past.

In a mathematical model, Einstein combines Time and Space into a single Continuum (measurement). The Continuum model describes *space and time as part of the same measurement rather as separate entities.* By combining time and space as a single measurement, Einstein was able to come up with a series of calculations to back his theory.

Figure 31: Spacetime

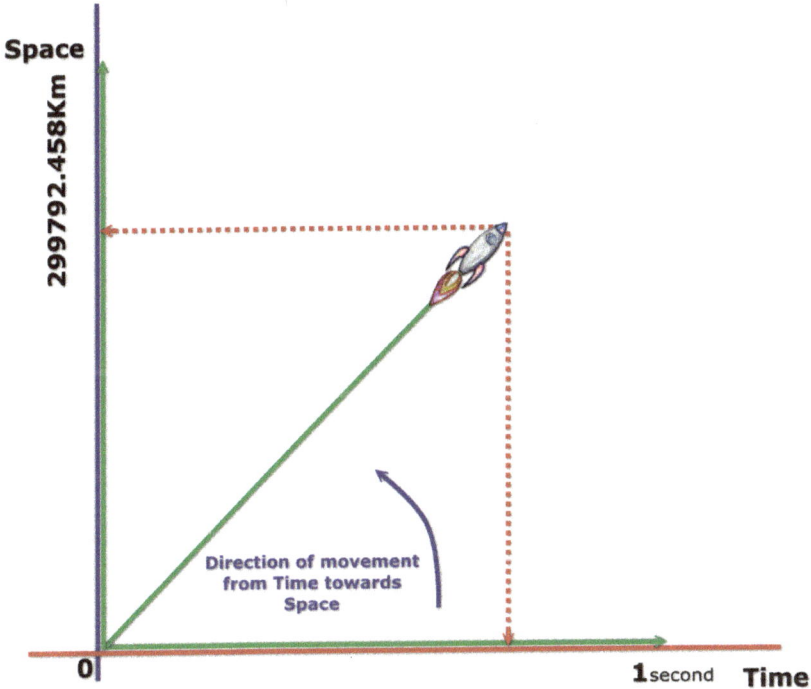

Time is universal across the entire universe. We have only one definition of Time, which applies to the entire universe. One unique and comprehensive definition covers and applies to the entire universe and everything it contains. Anything that happens across the universe happens in the same timeframe; we cannot have different time and time zones for different areas of the universe.

The Universe has, and is, only one time-zone and it should be seen as one uniform event, and only one Time definition can be applied to it as a whole. Of course, there are local times within certain areas where time can exist such as our solar system.

There are areas time cannot exist such as vast unoccupied areas or areas where time cannot be determined such as a group of stationary entities or entities not in formation and with no constant movement. In such cases, local time does not exist. Time is adapted and inherited from the universe.

Time's relation to the universe is in the shape of a pyramid in which Time is above all and applies to everything below. Time is a one-way adaption/ definition from the universe to all the entities in it. Entities, individual or collectively, cannot give or define Time for the Universe. However, the collective entities such as galaxies and systems can have own local time and set it for the entities in them. For example, we cannot create a Time Zone of Milky Way or any other galaxies or give time to the universe, but it can give Time to its peers and entities in it. The Milky Way has its own Local Time, and if it cannot be determined, it will adapt and inherit the Time from the Universe (*UniTime* or *UT*).

For example, the Earth has its local time, and it gives it to entities belonging to it, including the human race. The local time of the Earth cannot be given to the Solar System nor to the Milky Way or the Universe, but the solar system's local time can be given to the Earth or any other entities if their local time cannot be determined. Any events and activities in the universe should be registered first by universe time and then by local time because, as one universe, anything that happens across the universe, whether it is just a neighbouring system or a galaxy 10 billion light-years away, belongs to the universe first as it happened under its ceiling.

Figure 32: UniTime, Applied Order of Time Pyramid

In UniTime all entities have the same time, the time applies equally to all entities across the universe at the same level. The time we have here on Earth in UniTime is exactly the same time and same day on a planet 10 billion light-years away. Of course, the time is not going to be Jan 1st 6 AM across the universe as Jan 1st is local time. In a large city like New York, the local time applies to all equally across the city and cannot be set on a street-by-street basis. The time is inherited from the east coast, which is a local time, and the east coast has inherited its time from the continent, and the continent from the Earth, and the Earth from the solar system, ...

Time also can be determined by using Relativity. Time to an observer is related to the local area or the universe to which the observer is related. The time across the universe can be measured using light as Einstein has said in his Theory of Relativity, *"The speed of light in a vacuum is the same for all observers, regardless of their relative motion or of the motion of the light source."*

Using a unified Universe-Time, we can investigate and understand events more easily. Single time exists for all entities in the universe, and all entities are subjected to it regardless of local time, whether local time exists or not. UniTime can be calculated alongside local time when only one observer is in a local time. If two observers from two different local times try to use time in their communications, they have to use UniTime for better and easier understanding.

A starship, for example, 100 years from now, is carrying 50 travellers in a very distant area of space, and they cannot determine a local time and the preconfigured time device is not working either. Travellers need to communicate with the Earth-Station and inform them of their time and location. The only way to do this is for both sides to use UniTime; otherwise, they will not easily understand each other.

Time Dilation has no effect on UniTime, as the gravitational force is not strong enough to influence it. Time dilation is a phenomenon that happens in black holes due to the strong gravitational force slowing down time. This dilation of time is in local time only and has no effect on UniTime; use of UniTime will make redundant the dilation in local time and make it look like a small bump in the road.

The Spacetime theory works fine in a one-dimensional environment, but in a three-dimensional environment, it will not work because, in a 3-Dimensional environment, time and space are always parallel in relation to each other and moving from one to the other one will have no effect on the length of time or space. You cannot have time without length or length without time; these two are relative to each other.

Figure 33: Space and Time are Parallel

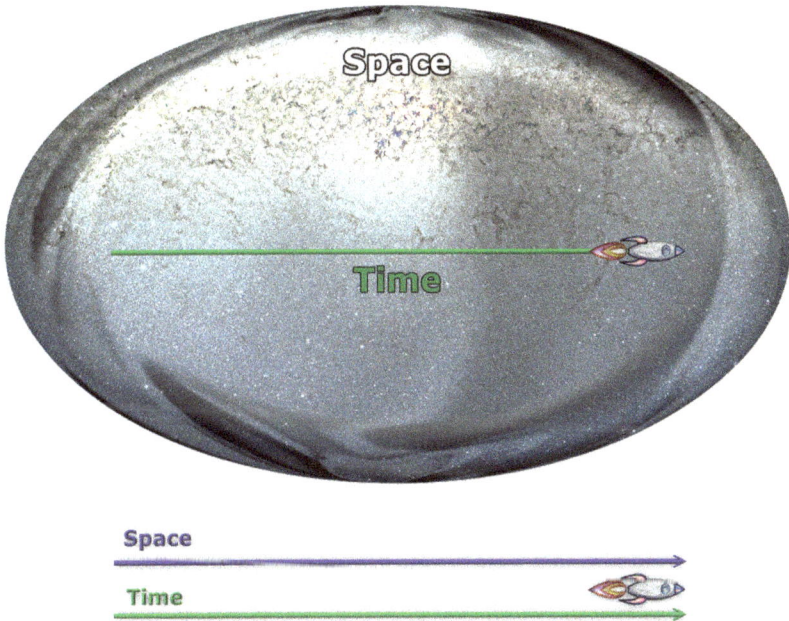

If a ship moves away from time and goes towards space, time will follow it and move with the ship. Any move by an observer will look different to other observers as it is said in the Theory of Relativity. A move from time towards space from one observer's point of view may look as though time is reduced and shortened, but to another observer a few miles away it continues looking unchanged. We cannot have two different results from one event. If Spacetime is accurate, then when an observer reaches a speed close to the speed of light and moves in a direction away from time and towards space, theoretically, time can stop completely. For one observer, the time has stopped, while to the other observer, only kilometres away; everything looks normal. Not only the time hasn't stopped, but also the position hasn't moved. Using two different points of view brings two different results for one event. The event itself has not been affected, as the evidence shows the observer on board the ship still travels at the same speed and direction as before. For example, if one observer sees a car explode from one angle, the second observer cannot see the car in good condition from another angle in the same space and time. An event always appears the same to two observers from different angles in the same space and time. If space or time changes, and, because of the change, the event changes, it continues to appear the same to both observers.

Spacetime appears to be just an optical illusion, more than anything else, even though there are calculations supporting it in a one-dimensional environment. If it were possible to reach speeds close to the speed of light, and move away from time and go closer towards space successfully, there wouldn't be any past to go to as we would have left it behind when we left time. We will end up in a time completely blank without any past events.

Remember the description of time we were told; how our daily time is like slides that constantly move back from us, it was just a way to illustrate time. Our lives are not like a series of slides; when we move away from time, these slides and past are left where we moved from, and the new place has no past to go back to. As we are stuck on 0 and time has stopped, we probably will have no future. We can't go back to the past because it doesn't exist, and we can't go forward in time because we are stuck on zero.

In Einstein's theory of Relativity "relative" to whom is important, relative to the observer on board the spaceship, the time has slowed down and will stop as soon as the ship moves from Time to Space; but relative to the second observer, the event is as usual, and nothing has changed. The speed of light is the same in the units of time. It is one event observed from two different angles, and it has two different outcomes, one event cannot have two different results in the same space and the same time. Applying the UniTime to the event will show no change in time and distance as the speed of the spaceship increases.

The theory of Spacetime hasn't made it clear if going back in time is sequential or if it can jump from one to another past randomly. Is it going one second into the past if we double our speed of light, and will it start going forward to the present time as soon as we take our foot off the gas? This is not clear, and the formula does not help either.

Time has no minimum and maximum speed. We use light speed as a measuring stick to measure time because the speed of light is the only event in the universe that is the same to all observers. The speed of light should not be seen as the pinnacle of time. The only difference between measuring a long distance using light speed per second and a tape measure is it takes longer with the tape than with light; otherwise, the method of measuring is the same and only the tools used are different. Light's speed is only a tool of measurement; not a means to go back to the past.

Maybe, if we are serious about wanting to travel in time, we should adopt a different tool and method instead of an optical speed. Maybe a 1981 DeLorean DMC-12 would be a better choice!

In Spacetime theory, the curvature of time is caused by the Earth's gravity as time has the role of a fourth dimension. It is not unusual for light to curve around some planets with larger gravitational force, due to the presence of energy in the universe. In the period of about 1 billion years before the Big Bang to about 100 million years after, there were a lot of large impacts and collisions between the planets and stars. These collisions released large amounts of energies, some of these energies were absorbed by other events/entities, and a large part was left unaffected and continued to travel across the universe. This energy, sometimes referred to as *Dark Energy* as its origin is unknown, is present, but low in energy force, throughout the universe.

This energy travels across like streams of rivers and in most cases in different directions. Due to the low level of force in this energy, it is impossible to detect with instruments we currently have. The presence of this energy will continue for billions of years to come, as its origins were billions of years past. It just starts reaching us and new collisions are being created as the result of stars exploding or collisions between galaxies. There are many collisions between stars, planets, and galaxies happening daily, and the energy released from these events continues to remain and travel in all directions across the universe.

When it comes to Dark Energy, I believe we need to talk plural about energies and not energy as singular. The origin and the timing of the creation of Dark Energies are different. The creation of the first Dark Energy was at least one billion years before the big bang, and the last of it was 100 million years after. The Dark Energy is a collective name for many events from many sources over a long period of time. These energies continue to be created on a much smaller scale because of the rarity, numbers, and the size of the entities involved in collisions.

The amount of the energy can increase or decrease at times, as it sometimes combines and creates a larger amount of concentrated force or the energy flows can cancel each other out when the streams of energy meet or cross paths. The direction of these streams can be influenced by the strong gravitation force of an entity. In some cases, the combination of an increase in the amount of concentrated energy streams and gravitation can influence the direction of light travelling in relation to the entity and cause an optical illusion that makes it appear to an observer that the space around the entity is curved.

In an experiment conducted by NASA in 2011, researchers claimed to confirm Einstein's theory of Spacetime. In this experiment, researchers sent a Gravity Probe equipped with sensitive instruments and gyroscopes to investigate the effects of Earth's gravity and time:

Time and space, according to Einstein's theories of relativity, are woven together, forming a four-dimensional fabric called "space-time." The mass of Earth dimples this fabric, much like a heavy person sitting in the middle of a trampoline. Gravity, says Einstein, is simply the motion of objects following the curvaceous lines of the dimple.

The result of this experiment was "The space-time around Earth appears to be distorted just as general relativity predicts," the experiment was a success.

In another experiment called "the Hafele–Keating experiment", testing the theory of relativity, in 1971 physicist Joseph C. Hafele and an astronomer Richard E. Keating placed three atomic clocks on board jetliners and flew them twice around the world eastwards and westwards. They compared the clocks against the fourth clock, and the three clocks showed time differences between them and the fourth clock. In this experiment, they confirmed that the time differences were consistent with the theory of relativity. The rate of change was 30 ×, or 300 quadrillionths of one second (0.000,000,000,000,3). Regardless of the narrowness of change, it remains a change in time and a confirmation of time dilation.

Many factors are involved with time dilation including gravity. These factors affect light's time of travel as light takes a longer distance to travel. When a ray of light is pulled by gravity around large masses or black holes, it travels a longer distance than it would in a straight line, and it takes longer for the light to travel this extra distance. If the distance from A to B is interrupted by C and the route has been extended by a certain amount, it should not be considered as an effect on time as it is an effect on space and the distance.

Because time is parallel to space and it follows space, clearly, when space is stretched and made longer so will time be. The speed of light remains the same, but the light has to travel more distance due to the curvature of distance/space and this curvature automatically translates to the curvature of time. The curvature of space around the Earth is not so much because of the gravity as the Earth's gravity is not strong enough to curve the space, but it is the gravity's impact on the presence of the streams of energy (dark energy) and the electromagnetic field that causes the curvature. Ultimately, it is the effect of gravity on a third party, which is amplified and has a stronger consequence.

There is a problem with Einstein's statement of *"space-time"* stated above. The statement appears to be contradicting itself. First, it describes *the mass of Earth dimples the fabric like a person sitting on a trampoline.* The dimple on the trampoline is caused by gravity in the first place. In other words: the gravity of Earth creates the curvature of the space around it and that, in turn, creates the curvature of time. Then gravity is *simply the motion of the objects following the curvaceous lines of the dimple.* It sounds like the question of the chicken and the egg and which came first.

The mass already has created the gravity and the curvature of the space is the result of the gravity. The second part of the statement cannot be correct, as the gravity already existed and is not the result of the following the curvaceous lines of the dimple. Is the mass, and gravity, the cause of the curvature of space? Or is the gravity the result of the motion of objects following the curved lines? Was it the chicken or the egg that was created first? At least we know the answer to this one—it was the egg! Or, was it the chicken? The egg, definitely.

The Earth travels at 1,674.33Km/h or 465m/s on its axis and at 30km/s orbital speed around the Sun. Similar to a travelling passenger jet, and the way it pushes the air, and the air around it is curved;

the Earth also travels in an environment in which, instead of air, it is surrounded by energy. The low-level stream of energy exists in space, and the energy is affected by the travelling planet's gravity and the electromagnetic field around it, which causes the energy to curve like the air around a jet. Would it be possible that the energy field is the fabric that dimples, not the time? Maybe the fourth dimension is not time and is the energy field we need to utilise to reach the conditions we must have to travel in time.

Time and Space are superglued together and cannot be separated. If the distance is extended so will time be extended, and this should not be translated into a method of time travel. The separation of these two will leave them both without any meaning and definition. Can we harness the energy just outside the Earth and use it to travel for free? Similar to the way a sailing boat uses the wind to push itself forward, maybe we can build a ship that harnesses and uses the energy to make interplanetary travel possible. We need to forget the past and look to the future. We need to see beyond the current time and tools we have and start building for the future.

More than 2,000 years ago, the Chinese used fire arrows to attack their enemy. Fire arrows were the earliest rockets we know to date. The Germans started using a liquid propellant rocket in the 1920s. The modern-day rockets are not much different from the earlier version, except they are a little more powerful and fuel efficient, we also get more grunt for our buck. Almost 70 years after the V-2 rocket we are still using the same technology to travel to space; we just use a larger size rocket. It is time we came up with better solutions and cheaper options to go beyond our earth.

To assess whether it is practical, let's put the idea of time travel to the test and understand how much energy is necessary to reach the speed of light. We will use, now, the retired Space Shuttle as an example. The Space Shuttle has a mass of 2,000,000kg with the booster rockets; and by using Einstein's equation

$E = MC^2$ to reach the speed of light at 300,000,000m/s (rounded up) we should get: $E = 2,000,000 \times 300,000,000^2$ the result is: 180,000,000,000,000,000,000,000,000 Joules of energy or Watts per second. It is equal to 640,000,000,000,000TWh (terawatt-hour). According to the theory of relativity, E is the amount of energy that is stored in the mass when it reaches the speed of light. This amount of energy is stored in the matter we use to advance forward.

The International Energy Agency IEA, in 2012, estimated the world's energy consumption was 155,505TWh for the year. It included the use of natural gas, oil, coal, and everything else. This makes the energy required for an hour to reach the speed of light using the Space Shuttle is equivalent to: $\frac{640\,000\,000\,000\,000}{155\,505} = 4\,115\,623\,291$ years. This is the total of annual energy consumption of everyone on Earth for 4.1 billion years.

We need this much energy for just one hour just to reach the speed of light, and this is what is said in Spacetime theory when time stops. If we want to move from this position and go back in time, I think we need a lot more energy after this! If we include the means of travel, including the engine, power plant, fuel, boosters, the structure that holds all these; the total mass will be much greater, and as a result, the energy required will be much greater. It is said the energy required to travel at the speed of light is infinite.

The Japanese nuclear power plant Kashiwazaki-Kariwa is the world's largest power station and could produce 8 GigaWatts of electricity per hour (8GW/h) before Japan's earthquake in 2011. Kashiwazaki-Kariwa is currently shut down for safety improvements. To get the space shuttle to reach the speed of light, we would need 514,452,911 of Kashiwazaki-Kariwa power plants to produce enough power for one hour. The Kashiwazaki-Kariwa construction cost was $20b, and it employed 4,000 staff. This would be equal to 1.028 trillion employees or about 147 billion times the total population on Earth. We don't need to say anything about the weight of the crew and the energy source.

Einstein in his theory of relativity has said an exponentially larger amount of energy is needed when the object moves faster. The faster it goes more energy it needs.

The calculation here is very basic, and it is just to show how much energy, approximately, this would need. To calculate it more accurately, the velocity also needs to be factored in:

$$E=ymc^2, \text{ where: } = \frac{1}{\sqrt{1-\frac{v^2}{c2}}} \text{ , } (v = \text{velocity})\}$$

Shuttle's velocity: $v = 27{,}869.5 km/h$

The calculation for the shuttle orbiter itself is:

Empty mass: 78,000kg; ($E = 78{,}000{\times}300{,}000{,}000^2$)

Result: 7,020,000,000,000,000,000,000 Jules or watts/s

Or: 25,272,000,000,000,000,000,000,000 w/h

Or:

25,272,000,000,000Tw/h $(\frac{25{,}272{,}000{,}000{,}000}{155{,}505}=162{,}515{,}674.7)$ years.

This is 162.5 million years of energy used on Earth by the entire population in order to have the vehicle travelling at the speed of light for just an hour.

Either the calculation is wrong or the formula to calculate the energy ($E=MC^2$), or it could be the guy who weighed the shuttle; because this shuttle needs infinite energy to reach the speed of light.

Maybe we are looking at this very much in the wrong way. We are going to use the technology that was invented in the 40s and 50s and then turned into a nuclear source to produce electricity. We are still looking at inserting Uranium rods into acid so it can get very hot, and boil the heavy water (^2H_2O or D_2O) and turn it to steam to run generators. The 60-year-old technology is not going to help us here, and the technology we have today is all about wind and solar and how to store the electricity in lithium batteries.

Perhaps we should come up with a new source of energy, based on the power of the atom, not so much Cold Fusion as it is just another form of current nuclear energy with less waste and almost no danger but heavily entangled in patent dispute; but something similar. Then again, if we had this kind of will to use cleaner energy, would we be burning brown coal? It appears greed rather than science drives the response to the need for clean energy.

Black Holes

A black hole is a region of space that has such extreme gravitational force that light, matter or radiation cannot escape from its field.

There are two types of black holes: Type-1 is structured by Type-0 particles, and Type-2 is made out of matter. Black holes are not always black and look like an empty space that sucks everything into it. There are some black holes with a star in the centre. The large star has extreme gravitational force, and it is surrounded by the usual particles and entities orbiting it. The stars are of the Type-1 category of black holes and are the leftovers from before the Big Bang that somehow did not become part of the events leading to the Big Bang and remained as they were. Studying these will give us more understanding of how the events were leading up to the Big Bang.

At present, the black holes are classified by size into four types:

- Supermassive black hole
- Intermediate-mass black hole
- Stellar black hole
- Micro black hole

This is unfortunate because it is the structure of the black holes that is important, not the size. The names sound more suitable for a list of sizes of coffee sold at a coffee shop. Every entity or complex formation in the universe has a size, and can, therefore, be placed in some category, but we should not categorise this formation by its size.

In the same manner, we first categorised planets and stars by their properties, then by their size, we should identify the black holes first by their composite structure, then their size. This kind of classification is similar to hiring a car when you are asked if you would like an SUV, Large sedan, mid-size, or small car. We should rethink the classifications of all entities, events and phenomena, or any other occurrence in the universe. The composite structure should be the first choice, and then other elements. This classification will help future generations to understand the universe better than we did before.

About 100 billion years before the Big Bang, when the extreme gravity was created, the gravity started pulling some more Type-0 particles from far away. These particles were left outside the recently created universe. Due to their distance from the rest, they did not become part of the formation of our universe when the condensation began. Now more than 200 billion light-years outside the universe, particles are pulled by the greatest gravity force the universe has ever experienced. They started moving towards the stars that were creating the gravitational force and at the same time, the stars were pulled by the gravitational force of bigger stars far off. These particles were towed by the gravitational force, which started them moving towards the newly created universe.

The energy that was put into the particles made it possible for the particles to continue moving even after more than 100 billion years, almost 1 billion years after the Big Bang, to arrive in this universe, where they are now. These particles were not part of the creation of our universe but joined it later. The particles still look like a cloud, not as large as they were originally, and later they entered a stream of energy generated by the collisions of stars and the final Big Bang travelling across the universe.

This energy stream, or energies, move like a ripple effect and can cause a superstorm-like environment after colliding with other energy streams. Particles carried by these streams can form a category five hurricane-like shape but with much stronger force and much larger size. As there is no atmosphere in space and nothing to stop them, they can rotate on their axis until something interrupts the rotation. The energy of the two collided energy streams combines and creates an even greater centrifugal force as it starts spinning around its axis. The clouds of particles create a vortex with a funnel. From a distance, it looks completely black, and it has less energy than the second type black hole. It cannot pull the nearby galaxies or light, but if light entered it, it could not escape. We cannot observe light emitted from behind, and if an entity, like a planet, entered the black hole, it would not escape and would be completely consumed and disintegrate into protons and electrons. The electrons and protons are at the top and outer edge of the black hole like a ring and can shine from a distance when observed. This type of black hole can be a smaller size than the other type, and it doesn't emit any X-Rays.

The other type of black hole is made mainly of matter with some small part being Type-0 particles. This black hole is created after two or more galaxies collide. The vectors of the galaxies are the opposite of each other. As a result, both galaxies consume each other, and the force of impact causes the two to create a vortex and become one event. The collision can take anything between 5,000 to millions of years to complete depending on the size of the galaxies. Our galaxy, Milky Way, is 100,000 light-years across, and IC 1101 is 2.8m light-years across. During the impact, a large number of stars and planets in the galaxies are completely destroyed, and the energy from those is added to the centrifugal force, making it even greater in power but smaller in size. If the planets and stars were not destroyed on impact, it would become a very large entity, but instead, it is a smaller size than the combination of the two galaxies, highly dense, and contains more energy.

Figure 34: Dust Disc around a Black Hole in Galaxy NGC 7052

After the impact is completed, the event acts like a tornado, spins fast, has a funnel, extreme gravitational force, and has the power to pull galaxies, and light. This black hole's structure is made of protons, electrons, neutrons, atoms, and molecules. It emits X-Rays and electromagnetic radiation. This black hole feeds its energy by consuming other galaxies and becomes even denser as the level of energy increases. The density of the black hole is not something we can create in a lab. I have seen pictures of electrons put together in IBM's lab to create the IBM logo; in similar ways, the density of the black hole cannot be greater than stacking electrons and protons in a large scale. There is a very little space between the spheres when placed together, and in some areas, the Type-0 particles fill this space. If the protons and electrons were flat, it could be denser.

The gravitational force of the biggest black holes is minimal to non-existent compared to the gravitational force that was initially created billions of years before the Big Bang. If time was, supposedly, to slow down in black holes it would have stopped completely and we would not have reached the grand finale of stage 2 of creation. Time, as we calculate it by using atoms and Quartz crystals, is slowed down only if the movement of electrons and subatomic activities are slowed down. If the subatomic activities are slowed down, then the binding of the molecules ceases to exist, and the molecules collapse. As we know, the molecules exist as the result of sharing subatomic particles. It can be said the extreme gravitational force can cause the cessation of molecules and entities

Imagine if we could attach a watch to a rope a billion kilometres long, throw it into a black hole, and pull it out a week later; will we see the watch shows some time behind ours? Not only won't we be able to pull the watch out, but also half of our rope is gone due to the cessation of its molecules. Nothing can survive inside a black hole. Not only is the gravity ultra-strong, but also the particles will go between the atom components and stop them from binding and functioning. Although black holes make a good subject for authors and Hollywood studios, they remain a mystery, as there are a number of unproven theories about them. Until we find reasonable evidence to show there are grounds for such theories we had better leave it as it is, just a black hole, for now.

To study black holes, perhaps we can create a mini version of a black hole in the lab. We need a ton of protons and a large industrial blender that can spin at high speed. The cylindrical blender should be made out of Perspex or glass so we can see inside. The protons spin at the highest speed we can achieve, and at this point, the contents will create a vortex, which resembles the real black hole, but it has no power of gravitation at all, due to its small size. The speed we can achieve in the lab is not near the speed of a real black hole, but it will show how heavier protons slowly end up at the top of the vortex and create a crust. This operation is not so much different from the production of Yellowcake.

Figure 35: Dense Protons, Electrons, and Other Particles

There is a limit to how dense the particles can become. Once they are placed tightly next to each other neither we, nature, nor black holes can press them for more density except by crushing them. Even the densest parts of a black hole can still have pockets of empty space in which particles don't sit as close to others. This limitation of density has a direct effect on the amount of gravitational force. When the density reaches its maximum, the gravity also reaches its maximum; the gravitational force is not infinite. A study of black holes will show the limitation of density and gravity. Possibly, we can reach the maximum density in a lab by simulating a black hole, and use it to study gravity, and how to create negative gravity. The amount of maximum gravitational force is directly related to the density, the quantity of particles in the dense area, the thickness of the wall of the vortex, and the amount of centrifugal force.

In Type-2 black holes, the density is directly related to the centrifugal force and the mass of the particles. The mass of the particles is limited and determined by the centrifugal force, which is limited to the speed of the particles and cannot be faster than the speed of light. If the particles in a black hole reach maximum speed approaching the speed of light, the centrifugal force causes the black hole to collapse, and the particles will disperse and go in different directions. This shows there is a limit that keeps the balance between the centrifugal force and the gravity inside a black hole. When a black hole consumes a star or a galaxy, the weight and density of the vortex increases as it contains more particles. It has a direct effect on gravity, and it increases the speed of the particles. The centrifugal force also has a limit: the further the particles on the outer side of the wall are pushed by the new particles from the inside, the less they are under gravitational force. The particles slow down relatively as they go farther from the wall but remain as part of the event. The thickness of the wall determines the maximum density and thus the gravitational force. It is the gravitational and the centrifugal forces that create the density. Gravity, on the one hand, pulls the particles and, on the other, the centrifugal force pushes them back; this creates a thick wall that makes the black hole.

The amount of maximum gravity can be calculated by knowing the amount of centrifugal force and the thickness of the vortex wall that relates to the speed of the particles. The wall's thickness has a limit, and it has a direct effect on the density that directly effects gravity. Although the amount of gravity is extremely high, it has its limitation and is not infinite. The amount gravity in black holes at the highest is still a fraction of what existed before the big bang.

If there was no limitation to the density, and thus gravity, in black holes, the black hole could become extremely large and powerful. They could completely consume the entire entities in the universe. The black holes cannot grow too big and become super black holes. They lose the same amount of particles as they consume. The black hole acts like a dismantling machine. At one end, galaxies enter, and are turned into particles, then, they are pushed out from the other end, like a coffee grinder. There is a limit to the size of the black holes, and to the quantity of particles they carry. This limit keeps them almost the same size until they complete their cycle, or are dispersed by other events, so the size of the black holes is also limited.

The centre of the black holes always stays the same, relatively, but the particles that once made the wall of the black hole are outside the event as they are pushed out by the new particles from the newly consumed galaxies. The former black hole particles don't go too far and stay around creating a cloud around the main event, sometimes creating a crust which can contain molecules and entities. From a distance, we may observe different sized black holes, but they remain similar in size at the core where all the force and particles are. It is the core of the vortex that has all the energy; the dust around is made out of particles ejected from the core.

Black holes do lose mass, and contract, but will gain mass and increase in size at the next event when a planet or galaxy is consumed. Black holes (both types) have a threshold and can hold mass and size at a certain level. The threshold of Type-2 is lower than the Type-1 due to centrifugal force, which pushes particles out of the black hole. It is possible to determine the age of a black hole by studying the amount of particle dust or thickness of the cloud and the empty space area around the black hole shaped like a disc. A larger disc and a larger dust cloud show an older and stronger black hole, and a smaller disc and smaller dust cloud show a younger and less powerful black hole.

It is a little bit different with stationary black holes. The Type-1 black holes are constructed from Type-0 particles and are mostly stationary. This type has a larger and much denser core containing Type-0 particles, and on the outside, it has a thick layer of protons and other particles. This layer is like a protective coating, and it keeps the core intact.

The gravitational force of these black holes is much stronger than the other type due to their larger size. They can increase in size, and it just further increases their gravity, so they continue to pull more entities. From the empty surrounding space around the black hole, it can be distinguished as Type-1. After billions of years, there will be nothing left around the black hole to pull and it remains unchanged until it dies, or it is destroyed by a very large galaxy crashing into it, but this is very unlikely. Even after the largest collision or death of a Type-1 black hole, the core remains intact, and it will never be destroyed. The core contains Type-0 particles which don't decay and also has a protective layer around it, made of protons and other matter, that will protect it from the most severe damaging forces. The core will survive and eventually will form a new black hole. The amount of gravitational force from this type of black hole is also limited and cannot be limitless and infinite although it is stronger than Type-2.

Type-1 black holes are identical to, or leftovers from, the formation of particles right at the beginning of the condensation, which formed this new universe. Studying this type will help to understand how our universe was originally formed.

Time and Black Holes

According to some theories in regards to time and black holes, the time slows down because of the strong gravitational force and a person can come back younger than his twin brother. We are told the gravity is strong enough to influence time in the many ways we measure time, and how it slows down rays of light. We have discussed how we measure time, by counting the pulse of microwaves generated by subatomic activities, by counting the 32,768 Hz of Quartz crystal, or by measuring the distance of 299,792.458Km it takes for light to travel in one second. If the gravity is strong enough to slow down these activities, then surely it can influence the subatomic activities of molecules in our body as well. If it does, then, because our body cells can bind and work at a certain level of subatomic speed, they will collapse, molecules will fail to remain bound, and, as a result, the whole structure will fall apart. It is not like body parts fall apart, it disintegrates into molecules and atoms instantly: it evaporates.

Gravity slows down the rays of light if they are going out of the gravity field. By measuring this, we will find it will take longer for the light beam to reach the 299,792.458Km in one second (perhaps it will take 1.5 or 2 seconds of Earthly time), but it will take a shorter time for the light beams from outer space to arrive at the black hole as the gravity is pulling them faster. This shows we cannot use light beams as a measure of time in events like black holes. Other devices we have will not work as intended, and we will need to find a way to measure time, different from any we have now. By comparing the two speeds, we can measure the amount of influence of the force of gravity.

Some may say we don't need to be on the black hole to experience the time-shift; we can be on a planet on the event horizon and experience it, at least, to some level. No! If we are on a planet on the event horizon, and the force of gravity is not affecting the planet's surroundings, the atmosphere or the cells and the molecules in our body, then the gravity is not strong enough to slow down the light or the time measuring device. The amount of gravitational force must be much higher in order to influence time. At this level, it is well above our strength to withstand such force. Our molecules will fail long before the gravity starts having any effect on light.

What if we could land a small craft on one of the comets that pass our planet every 100 years, and have the craft carry a small atomic time device? The comet and the craft will soon leave the solar system and go to areas without any gravitational forces. At the same time, we send another craft to Jupiter, which has gravity 2.5 times that of the Earth. After about 50 years when the craft continues to broadcast the time, perhaps it will show a difference between the times in higher gravity and no gravity conditions.

For some reason, we have always looked into the effect of gravity, and how it slows down light, and interpreted it as the slowing of time. This event is in two parts, and we should look into the first part, too. The first part is when the light approaches the black hole, and the gravity pulls the light so that it increases the speed by the same amount as it is slowed when it leaves the black hole later. The amount of gravity is the same on both occasions, and the rate of its application is also the same. When light approaches the black hole and the speed increases, this means the time becomes shorter and faster. The light travels faster in the first part of the event and slower in the second part, looking at the whole event as one, light travels at the same speed and time is unaffected. Applying the UniTime to the event will show the black hole, although it has some effects on local time, has no effect on UniTime in its totality.

CHAPTER 9:
THE CREATION BEGINS

Event Correlation

Back to the armchair position and we can witness the unfolding events. The time starts and ends are becoming further apart and it means we have time for longer periods. The existence of time can last for at least hundreds of millions of years, but still, it is not a permanent creation of time. The ability to observe and calculate time comes from a routine and constant movement of entities in the area. If we could see in complete darkness, we could witness the entities orbit and follow the same path for millions of years before they are interrupted by a passing object and the collision that will end their time and sometimes completely ends time. As we can see in the creation timeline graph, light has not been created yet, and its creation is billions of years away; but we are able to calculate time, based on the constant and routine movement of entities. To have time, all we need is a constant, and the constant is here, the existence of time has nothing to do with the presence of light. Time existed hundreds of billions of years before the creation of light.

The interval of time continues for at least another trillion years while the creation continues to go forward. Time continues to be created and to end, ranging from a few days to a hundred billion years or so. These starts and ends are not connected to the time we have now. The start of the creation of time is the time that connects us to it. Time will not end until this universe ends. The creation of our time is the last one that still connects us to it. The entities are now larger and have stronger gravitational force. They are constructed like a planet with all the particles stuck together, the force of the gravity is creating a core which is extremely condensed and is growing in size. The entities continue to collide, and every time they collide, it takes decades, and sometimes centuries, for the particles to regroup and create a larger entity.

No event, in particular, happens for next trillion years, except the entities are becoming larger and the force of gravity becomes greater. The cores of these entities are now larger than average planets we now have in our solar system. The planet-size core with more particles surrounding it makes the entities larger than this universe has ever seen. The collisions become harder and heavier creating greater destruction. At the same time as the larger entities are formed, more entities are destroyed and go back to the clouds of particles they once were. The dispersed clouds of particles won't remain as clouds for too long as the existence of large gravitation attracts them; they are regrouped and create new larger entities. This process continues, and large gravitational forces are created. The large gravity pulls more particles from far off and pulls other entities towards itself. It is, at least, 10 trillion years' gap between the creation of gravity and the creation of large gravitational forces.

The constant collapse of large entities slows the creation process. The creation is still going forward but not as fast as it could. The reason for this is that the entities are not like planets made from molecules and atoms. Nothing holds these giant objects together, they are just creating a shape, which is not in any particular order, and just looks extremely large. If you made a paper plane and threw it at the entity, the paper plane could go through it for a considerable distance before the particles dismantled the molecules.

Figure 36: The Creation Timeline

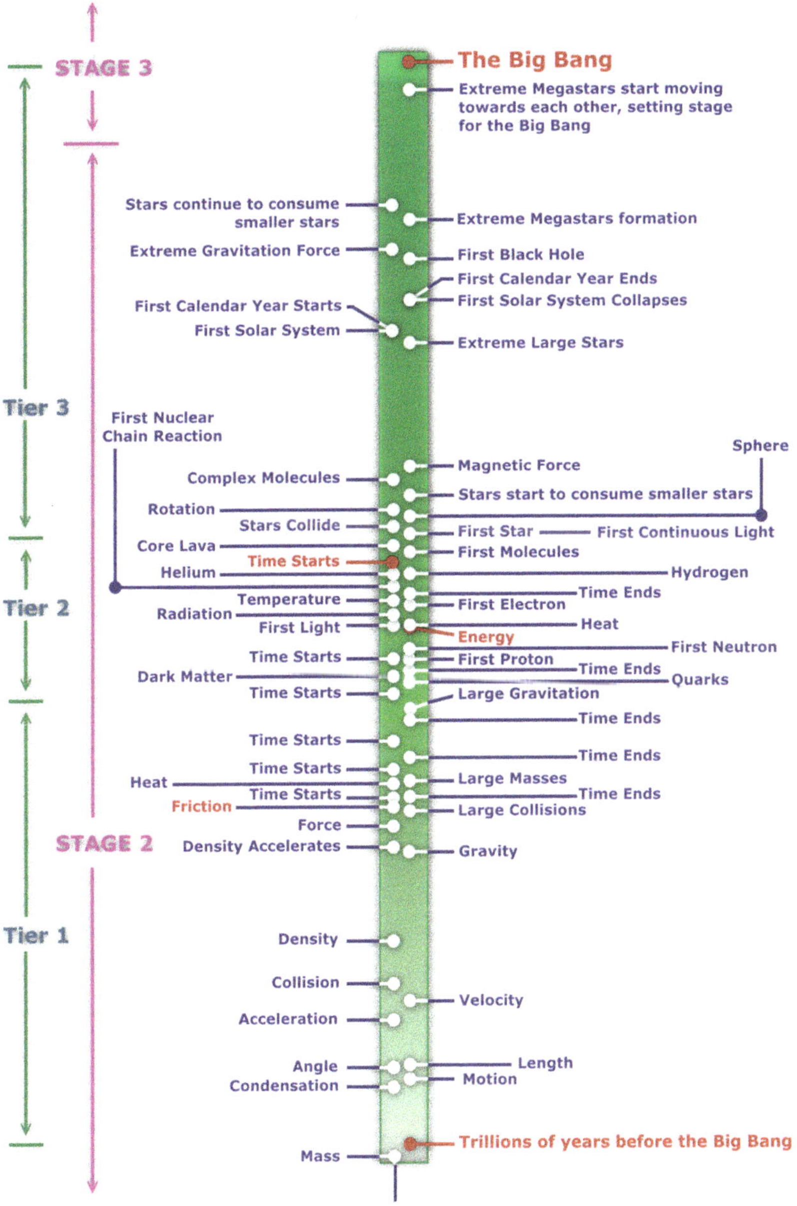

The large gravitation continues to create much larger entities for billions of years. The gaps between the clouds of particles are now hundreds of thousands of kilometres wide. Although a large portion of the clouds of particles joins to the objects and forms the large entities, there are still immeasurable amounts around, and it will take trillions of years for the particles to join the others and create larger entities. The number of entities that exist at this time of creation is high, but not enough to take creation to the next stage. This is almost the halfway-point from the time condensation began to the time the big bang started.

The First Proton

There is no such a thing as the first proton. It simply can't be. Protons, or any other particles and subatomic particles, are not something made in a large production line in a factory, with a big press stamping protons out of a sheet, or an injection moulding machine making them in a die and pushing them out onto a conveyor belt, where they are packed and sent all over the universe. The protons are created in amounts of hundreds of millions of cubic kilometres at a time.

Figure 37: Structure of an Atom

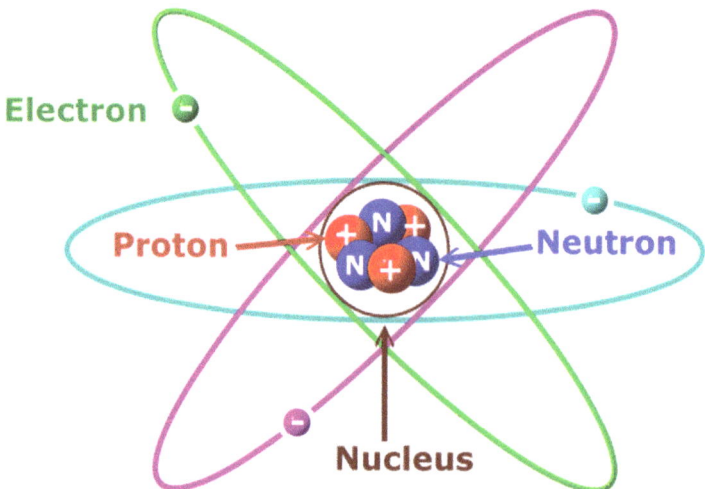

Making protons is a very long process, and it can take more than a trillion years to complete. The temperature is still around −8k°C as this is an essential condition for the creation of protons and other subatomic particles. It would have been impossible for the protons to form if the temperature was any warmer. From this point forward, the subatomic particles are created in an absolute ZPE environment. The creation of energy is not too far away, and the creation of all the new particles is taking place in the Zero-Point Energy state. This is a true ZPE environment, which existed all across the universe. These conditions will never happen again. ZPE can and will exist in the future but locally and not across the entire universe. The status of ZPE across the universe ended once energy was created.

In 1845, a British physicist, William Prout, theorised the structure of atoms and the proton. The name was given to the hydrogen nucleus by Ernest Rutherford in 1920. A proton is a positively charged stable subatomic particle that, together with neutrons, makes up an atomic nucleus. A proton is now a Hadron in the modern standard model of physics and is composed of three quarks: two up quarks and one down quark.

Protons are not like a hollow sphere with three quarks inside. The scientists have discovered the protons are made out of three quarks, and it does not imply they are inside a sphere. The three quarks are bound together by the powerful force between them and exchange electric charge. The Elementary Charge is the electric charge carried by a single proton.

The protons were created in a very specific way. The entities being created by Type-0 particles are now the largest in the universe, and their numbers are growing. At this time of creation, about 20 trillion years before the big bang, the entities continue to collide and create even larger entities. The masses of the entities have created great gravitational force, and it pulls the entities from far away and creates more large objects. At this time of creation, the cores of the entities are much greater than planets, and they are becoming even denser every time they collide into each other.

The heavy collisions between the entities cause the cores to bind to each other; the collisions take months to complete. The leftovers are disbursed into space and are attracted back to the newly larger entities or the force of the collision pushes them far enough that gravity won't be able to pull them back and they will eventually join other entities.

The binding between the cores starts to create a new particle. The particles are created in an absolute energy-less condition. By the force of the impact, the Type-0 particles are forced into each other, and a new particle is created consisting of two or more Type-0 particles. The new particles continue to make the cores of the entities alongside the Type-0 particles. This continues for a few trillions of years. The entities are now mostly made out of the new particles. The cores are now larger and looking more like the planets that we have now. These planets continue to collide, and every time, new particles are created.

The new particles at the core of the planets add more to the gravity due to the fact they increase density. The small gaps between the particles are reduced to zero, and the stronger gravity puts more pressure in the cores of entities. The planets/entities are colliding into each other with much greater force, and more resistance is present. The collisions are now creating solid entities. These entities are now harder than rock on the surface and even harder and denser in the core.

After trillions of years, the universe has become an empty looking place. Once there were clouds of particles, now turned into entities larger than planets. From a distance, if we could see in absolute darkness, it appears the universe is full of rocks of many shapes. The entities are composed of Type-0 and the new particles.

Slowly, over hundreds of billions of years, the entities greater in size start creating even greater gravity and greater collisions. Large gravitational force is bringing the entities closer and creating collisions on a scale that never happened before. These collisions are taking place daily, but take at least 100 years to complete due to the size of the entities and the force of the collisions. After every collision, the new core is made for the newly created entity and, over time, the core is made more and more out of newly created particles. The Type-0 particles take longer to regroup and come back to the new core due to their lighter weight and being less affected by gravity. Every time there is a collision, many Type-0 particles are dispersed like dust. Most go back and join the new entity, and the rest continue moving in the direction they started in, when they were forced into space, and eventually join other entities. On every occasion this happens, the Type-0 particles end up the last particles on the surface of the new entities and the new particles are at the core of them.

These collisions and re-creations continue for trillions of years until entities are made completely out of new particles. The process of creating planet size entities continues, and at the same time the entities composed of new particles are created, more entities composed of Type-0 particles are created too. This continues and is a long process.

One of the collisions between the two entities composed of new particles (we should call it a quark now), causes the cores to be pushed into each other by so much force that the force of the impact causes the particles to bind and create a new particle. The new particle is called a proton.

Figure 38: First Proton

At first, the protons are very rough looking with a very rough surface. Just as peppercorns are ground, the protons are ground in the collision and very quickly lose their roughness and become smoother and shinier.

Figure 39: Four Stages of Creation of a Proton

The creation of a proton and the smoothing of its surface occur in one single process. The rough proton does not remain in that state for millions of years and then slowly become smooth. It happens in one single pass. Cubic kilometres of protons are created at once and in a few of seconds or minutes. The new particles are made from a larger number of quarks than three. The cold forging of the new particles stores the new properties and new abilities in the particles.

At this time, these particles do not have any properties or any abilities. The original colour of the proton is black, as the colour is determined by the structure of the molecules and the composition of an entity. A proton, as a single particle, has no structure and composition, and therefore no colour. Because light cannot bounce back and be reflected by it, we cannot see colour. We see an object's colour because the object absorbs all other parts of the spectrum of light and reflects only the one determined by its molecular structure. The finished surface of a proton is so smooth and shiny that it reflects all the light, so it appears white, but is still black and it is only because of the shininess of the surface that it reflects light.

The process of creation of particles, quarks and protons, is not something that once it starts will continue the creation like a chain reaction. The process can stop any minute, and the start of the next one can be thousands of years later. The particles are created in individual sessions. Every time there is a collision, the creation of particles is not certain. Not every hit and collision is a success and creates protons, sometimes it creates quarks, and sometimes nothing happens, but the particles regroup after the collisions and create a new entity, which eventually will become the entity that creates protons. The conditions must be right for the creation; things like the size of the entities, and the amount of force they produce; the core of the entities and the density are also main factors in creation. Not every large entity, and not every hit, is going to create protons. The creation of protons is a rare event; it is at the early stage of the creation of subatomic particles, and it will take trillions of years to complete.

If we look at the long timeline of creation, we are able to see the clouds of particles slowly come together and create objects, and the objects become entities and grow in size, becoming great entities. The entities become highly dense; they collide and create new particles called quarks. The quarks follow the same process, join together, and create large masses and entities. The large masses of quarks become extremely dense due to the gravitational force and create larger and denser entities. The entities, over the years, start colliding and very rarely create new particles called protons.

It very much sounds like a production line of a factory. The only thing missing is Charlie Chaplin and his big spanner.

Protons are composites of quarks forged in extremely cold conditions. The cold conditions allowed the quarks to create their own properties and these are passed on to the proton. The forging and binding of the quarks removes any need for energy to play the role of catalyst, and the quarks are transformed into a new particle without any external influence. Later, when the temperature rises, the binding remains and will produce a strong energy as the result of the cold binding.

Many of the collisions fail as they don't create protons, but a small number of quarks still bind and make a kind of particles that are not quite like anything else. These new particles are a combination of Type-0 and quark particles; they don't have properties and won't take part in creating protons in the future. They are a kind of half-done particles. These particles, known as Dark Matter, will eventually, when there is a larger number of them, create other particles called Neutrons. The Dark Matter particles are created in a similar manner to the quarks but are slightly different. The quarks are made out of Type-0 particles only. The Dark Matter particles are made out of a mixture of quarks and Type-0 particles. These particles will create electrons after more than 500–1000 billion years or more. The amount of Dark Matter at this stage of creation is just not enough to allow the creation of the other subatomic particles. Type-0, quarks, and Dark Matter cannot emit any radiation due to the structure of the particles; any particles created after these can emit radiation.

The difference between the particles is the number of Type-0 particles that are mixed with quarks and the Dark Matter. At the very early stage of creation, there are only two kinds of particles, Type-0 creates the protons, and the quarks are in fact a failed creation of protons. These particles continue to regroup, repeatedly, to create giant entities, which can take hundreds of millions to a billion years. These entities are now composed of few particles, Type-0, quarks, and Dark Matter. Dark Matter is also a by-product of failed proton creation, a mixture of Type-0 and quarks created the Dark Matter in separate events. If the creation was perfect, and the number of particles was right at the time of creation, the event would have produced a fully working new particle like the electron. As the number of particles was not enough, and the mixture of different types of particles wasn't suitable, electrons were not created at this time. The creation of new particles was delayed for trillions of years until the right mixture of particles was present. This cycle continued for another trillion years, at least.

As we witness the events at this point in creation, the conditions are unique and complex. The temperature is still at the lowest the universe has ever experienced, and the Type-0 particles are in the most condensed form. The impact of collisions, the great force of the impacts on the particles, and the cold forging of particles into quarks and Dark Matter can never be recreated, even by the universe itself. Certainly, the creation is still, after nearly 1000s of trillions of years, making Type-0 particles at some remote area of the Greater Universe. However, when the particles are processed and turned into other particles such as quarks, Dark Matter, protons and the rest, the universe cannot recreate the Type-0 particles in the new conditions. The universe is changed totally and absolutely, and the conditions are not there anymore. This is only a one-off event, the speed of the process is extremely slow, but the slow timing allows the creation to complete. If the creation had taken a fast route, the creation and transformation of particles would not have happened, and it would have been the greatest fail of epic proportion. The creation is in no hurry, and it is taking its time.

Due to the conditions required for the creation of Type-0 particles, these particles never had any properties, but when they turned into quarks under sudden and extreme pressure, the properties were given to the quarks; or the quarks adopted the new properties, and it is the quarks that make the protons. The properties of the quarks now make up the properties of protons. These are not only modified from the properties of quarks, but more properties are added for the new particle. By smashing the protons, we cannot remove the properties or convert them back. The properties remain intact in the smaller size particles. It is like smashing a nut, the nut pieces are still nuts, there is no definition of the size, and how big it has to be to be called a nut or if it is smaller than a certain size it is not called a nut anymore. Even if we were able to grind protons, the proton powder would still have the proton's properties or at least some of them. The proton powder also will never be converted to Type-0 particles; reverse engineering will not work even at this level.

The Oxford dictionary defines a proton as:

A stable subatomic particle occurring in all atomic nuclei, with a positive electric charge equal in magnitude to that of an electron.

The Merriam-Webster dictionary, in regards to the mass:

... an elementary particle that is identical with the nucleus of the hydrogen atom, that along with neutrons is a constituent of all other atomic nuclei, that carries a positive charge numerically equal to the charge of an electron, and that has a mass of 1.673×10^{-24} gram...

Protons cannot <u>possibly</u> be made out of only three quarks; three quarks give the proton the ability to spin and positive electric charge, but the size of three quarks together is smaller than the proton.

Protons are composed of a mixture of quarks, Dark Matter, and Type-0 particles. If the protons are made of the three quarks, then by dividing the mass of proton (1.673×10^{-24} *gramme*) by three should give us the mass of one quark,

$$(Quark\ Mass = \frac{1.673 \times 10 - 24}{3}),\ \text{or } 5.576 \times 10^{-26}$$

right? ... Wrong!

In regards to the size of the protons, the Oxford dictionary has this to say:

> The mass of the proton is 1,836 times greater than that of the electron.

So what are electrons made of if the size of them is smaller than one quark?

$$\{\frac{1.673}{1836} - 9.11 \times 10^{-29}\},$$

and the result is disappointing:

$$\{(9.11 \times 10^{-29} \neq 5.576 \times 10^{-26})\}.$$

The numbers don't add up, and it appears the protons have a higher number of quarks than is said, and dividing the total mass of a proton by three is not correct. If the proton is 1,836 times larger than the electron, and the electrons are also made from quarks, then the protons either must have been made out of at least 3,672 quarks (at *least* 3,672 if we say the electrons are made out of at least 2 quarks), or the electrons are made out of something smaller than quarks.

The protons are not made out of 3,672 quarks, it is much less than this, and it takes more than two quarks to make the electrons. Both protons and the electrons, and the other particles are made of a mixture of quarks, Dark Matter, and Type-0 particles forged under extreme pressure at extremely low temperature. *What makes the particles different from each other is the multiplier number of the three-particle mixture that creates the composition of the subatomic elements.* All particles are solid not hollow and extremely resilient, there can't be any other particles lose inside them moving around.

By smashing protons, we will not get the quarks or any other primary particles, and by smashing quarks, we cannot get Type-0 particles. The creation cannot be reverse engineered. Smashing the protons will give us smaller proton pieces which still carry the properties of the proton, even though the smashed pieces are smaller than quarks, they are still <u>not</u> "God's Particles".

We cannot recreate quark particles in a lab, nor can the Universe itself.

At this time of creation, more than 5 trillion years from the creation of protons, large number of entities are composed mainly of protons. In one of the collisions between the giant entities, the collision takes a long time to complete due to the large size of the objects and slow but extremely forceful collision. The extreme pressure of contact between the two objects, which is friction, creates a very large amount of heat. The heat has been present in recent times when collisions occurred; and due to the extreme cold conditions, it has never been a great indicator of the presence of energy. This time it is different: the heat rose to a very high temperature in a large area of the collided entities, and it remained high for the duration of the collision, which took months to complete. There was no fire or anything, the friction just generated a large amount of heat, and the protons became very red. The result was a sudden creation of ENERGY.

First Energy

Energy is the property of an object that can be transferred to other objects or be converted to other forms.

Until this time of creation, none of the particles had properties. Once the quarks were created and the protons from the quarks, both particles gained properties from the creation. The amount of force the friction put into the Type-0 particles, and transformed them into a new entity, gave the particles properties as well. The properties of the particles were not obvious and readily available to the particles, but the force of the friction created a large amount of heat and released the properties of the particles, which turned them into hot, red and active objects. The objects became active only momentarily, very quickly the heat disappeared, and the cold temperature turned the active particles into a ZPE state. This time it is different, the particles came to life from completely dead particles, and when they were exposed to energy they became active particles, then they go back to the ZPE state as a particle with changed properties. This kind of particle is totally different from the one it once was; this particle can become active at a much lower temperature than the old one. The same particle in the previous state needed a higher temperature to get to the point where it could be active and properties can affect its behaviour; now the properties are stored in it, the cold temperature just puts it into sleep mode until next heat-wave brings it to life.

The result of this sudden creation of energy is an immediate release of a new particle called a Photon. The Oxford dictionary's definition of a Photon is:

A particle representing a quantum of light or other electromagnetic radiation. A photon carries energy proportional to the radiation frequency but has zero rest mass.

Photons are packets of energy in the form of electromagnetic radiation; they have no rest mass or electrical charge. The light is the result of radiation generated by the heat and does not last long. The amount of radiation is so small it cannot be considered as a new creation. The bright light briefly lights up space and quickly disappears into the particles. The heat also quickly disappears in the cold environment. The warmth stays between the entities for a short while.

The collisions continue to dominate the universe and cause the entities to create heat; sometimes flashes of light can be seen. The reason the collisions occur very regularly is the close distance between the entities. The Type-0 particles at the very early stage of the condensation came from a vast distance and created the entities, and now the entities are brought together by the large gravitational force to a much smaller area of the universe. This area is where all the entities gathered, and there is a great distance to the edge of the universe which this universe was once a part. From the outskirts of this gathering of the entities to the line that separates our universe from the great universe is trillions of light-years of absolutely empty space. This area is a perfect example of the definition of a vacuum environment. This emptiness will not last for too long, the big bang will push the entities back and will fill the space once again, this time with different kinds of particles.

The energy created in recent events is not only in the form of heat but also is in other forms as well. There are many forms of energy, they include:

Kinetic, Potential, Gravitational, Mechanical. Electric, Nuclear, Magnetic, Thermal, Heat, Radiation, ...

We did have Gravitational energy at the very early stages of creation and it always comes with kinetic and potential energies. The gravitational force had nothing to do with the creation at that moment, and the energy was considered irrelevant to the earliest stages of creation. The gravitational energy and the two of its siblings—kinetic and potential energies, could not have kick-started the creation as friction does at this time. Only the events that led to the creation and had a key role were important as a creator. The kinetic and potential energies were microscopic at the time of creation of gravitational force and did not play a crucial role in creation. We discuss only the events that changed the direction of creation or played a role that led to a greater event of creation.

The energies created and released by the collisions created heat and, over time, created mechanical waves. The particles were not bound together strongly, and the mechanical waves caused them to act like waves in an ocean and sometimes carry the heat with them. Radiation energy was also created and released by the collisions, and it transferred the energy to areas of space where the mechanical waves or heat couldn't reach.

The Oxford dictionary defines Radiation as:

> The emission of energy as electromagnetic waves or as moving subatomic particles, especially high-energy particles, which cause ionisation.

At this time of creation electromagnetic force and the ion have not been created so moving subatomic particles were the only part that applied to the current environment.

The energy created as the result of friction is constantly radiated to the other particles. Some of this energy is transformed into heat and the rest just creates waves, travels through space, and has an impact on the particles. The amount of the energy is increased every time a collision occurs.

The tidal waves of energy travel across the universe and push the newly created particles tens of thousands of light-years away and into places where they haven't been. The new particles mix with the old particles and join to create large objects at the other side of the universe. It will take quite some time for the regeneration of the particles to spread. The creation of quarks and Dark Matter is not happening from one source anymore, those that have been created are pushed like a constant sandstorm tens and hundreds of thousands of light-years to other territories. The areas that were behind the creation of new particles are receiving new particles that will mix with others and create other new particles. The whole universe did not reach the same stages, and enter a new era, at the same time. Some arrived at a point of creation earlier than other areas and started creating particles that could only be created at that point and nowhere else. With the energy created in some areas and the radiation pushing the new particles to old areas, there is a new opportunity for the creation to take a different path and create new particles.

Until now, without any energy or the existence of any other particles, the creation could only create one or two new particles at a time. But the new particles have travelled, sometimes, millions of light-years and arrived in the old areas; they give a chance to the creation to create completely new particles for the first time. Going through the same process as the creation went through in the past 5–6 trillion years, it has a chance to create something different with the same process. It is, in a way, similar to a production line of a factory; some of the products from the end of the line are put back at the start and mixed with primary ingredients to make a different product. Unlike a factory, the creation cannot continue this for eternity, and it has to end soon. The creation will create all the electrons, protons, and other particles and eventually will run out of particles. There is a limited supply of particles in this universe. It was limited when the particles started separating from the rest of the ocean of particles at the very early stage of the condensation era. This universe is left with all the particles it needs to create; all the entities we see today, and many more were destroyed before and during the big bang.

Although the number of particles is extremely large, it is not an infinite supply. The means or action that produced the particles in the first place definitely still continues to produce particles at the Greater Universe level. However, those particles cannot come to this part of the universe anymore; after nearly 1000s of trillions of years the conditions are changed, and particles cannot travel to this part (not that the particles ever travelled anywhere in the first place; energy is required for particles to move and at the stage particles are created, there is no energy to move them).

We could, if we had the means, count the particles and put a number on the total amount. It would be an insane number, so large we couldn't even see the end of it, perhaps as long as a few return trips to Mars if written in a small font; but, still is a number and we could put a total on it if we had the means to count them.

It is not too difficult to calculate the number of particles in a different way, and the result won't be a large number as long as a return trip to Mars. The original size of the current universe, if we say it was 10,000 trillion light-years across before the condensation began, and the same on Y and Z axis; it makes 1,000,000,000,000 trillion cubic light-years or 10 Quindecillion (10^{48}) cubic light-years. The number of particles in a cubic metre multiplied by the size of the universe will give us a rough estimate. Still, it will be a ridiculously large number.

This can keep you from sleeping: one cubic meter of Uranium 235 is 19.05 tonne. Uranium has a half-life of 703.8 million years, 1 kilo of U235 has 2.56×10^{24} atoms, and every atom has 92 protons, 143 neutrons plus the number of isotopes and electrons. You do the maths and find out how many protons in 10 Quindecillion cubic light-years.

The limited number of Type-0 particles can only make a certain number of quarks and other particles, and, as soon as they are made, the universe will end its creation process and get on with processing the new particles in different ways. The total number of particles will be fewer as they are combined to make the protons, electrons and other particles. The total number of the new particles (protons, electrons, ...) will also be reduced, in turns, to make molecules, and the molecules will create entities that will shape the universe we see today. The bad news is, the universe is not finished with its creation and is between stages. This quiet time we have now, since the creation of our planet and everything on it, is just a break between the stages. Our Solar System is part of the Milky Way Galaxy, which is part of the Virgo Supercluster, and it will move and crash into other Galaxies and clusters eventually, and end, as has happened many times before. This is very similar to the collision that created Virgo billions of years earlier. What we have now, a quiet corner of the universe, is only temporary and will come to an end. The good news is, you and I won't be around to see it.

This time the creation is quicker than before and it takes only a few hundreds of billions of years, no more than a trillion years, to create a new particle very far from the place protons were created. The new particle is created in the same way as the other particles, but the ingredients are different this time. The energy has pushed a lot of quarks and Dark Matter to the far corner of the universe where the creation is still taking place with the old style of process. This area of the universe may be only a few hundred billion light-years away, but in the creation timeline it is at least 5–6 trillion years, and sometimes more, after the creation.

On the one hand, the creation has taken 5 trillion years to create some particles, and, on the other hand, new particles are introduced to the process, and the timeline in this area has changed. Until now, the timeline was the same across the universe, but now suddenly the same process that was taking 5 trillion years to create a particle, will take only one or less.

We cannot have two different timelines for the creation; the universe has only one timeline, which is called UniTime. The timeline needs to be adjusted to accommodate the different time duration for the creation of particles. UniTime will allow us to look at the whole event as one period.

The creation continues, and it creates a new particle after almost 6 trillion years from the creation of the proton. The new particle is called an electron.

The First Electron

An electron is an elementary subatomic particle with negative elementary electric charge. An electron has a mass of $1/1,836$ of a proton, and 2 electrons cannot occupy the same quantum state simultaneously due to their properties.

Unlike protons, the electrons are made of more kinds of particles but lesser numbers. The particles are Type-0, quarks, Dark Matter, and a few proton particles. The proton particles are not quite complete protons: these particles did not reach the final stage of creation to become protons and were left unfinished, similar to Dark Matter. This extra number of proton particles in the electrons allows the particle to adopt a different property and become a different particle that interacts with a proton to create an atom when the time is right. All of these particles share some of the particles, some more and some less than the others. All particles have the Type-0 as the main composition, with the quark and Dark Matter to complete the structure of the particle. Some, like the electrons, have a small amount of incomplete protons or pieces of protons in their composition. There may well be other particles that have partial particles of the proton and the electron in their composition and fewer quarks and Dark Matter. These particles were created trillions of years after the creation of protons and electrons, or even neutrons. They can hold the properties of the two particles at the same time. As was said earlier, it is the multiplier of the mixture of the different particles in the subatomic elements that makes them different from each other and gives each particle different properties. The amount of different elementary particles, Type-0, quark, and Dark Matter, results in the creation of different particles. All particles have the first three elementary particles in their composition, but it is the multiplier number of the mixture and the additional elementary particles, such as failed proton parts, in the electrons that give them their differences.

It is the multiplier of the mixture and the ratio of the particles that allows the creation of the subatomic particles. The limited number of combinations and the ratio of particles allows only a certain number of subatomic particles and not a larger number of combinations because of the limitation of the multiplier. Only three elementary particles created the proton. First, there was only Type-0, and then quarks and Dark Matter were created from Type-0. Finally, the combination of the three particles' and the mixture ratio created the proton. The multiplier was 3 for about 5–6 trillions of years but changed to 4 after the creation of protons. As you know, only a very small number of combinations can be made from only three elements, and only one type of new particle was created.

From the creation of the new particle, there was a large amount of failed creation of particles, which produced a fourth element in the creation and the multiplier became four particles in the mixture. From the creation's path, it is entirely possible that the electrons were not completely identical and some failed, the failed electron particles became the fifth element in the multiplier and created new particles such as photons or antiquarks. The multiplier at this time of creation has reached 5 and will not go any further, the creation has reached its limits. The creation of new base particles will stop shortly.

In about 2–4 trillion years from the creation of the electrons, the creation will stop creating particles; this will end the creation of subatomic particles. From this moment, the creation cannot create any new particles as it did previously. The creation continues to create new entities from the particles, but not any new particles. We are not going to see any further increase in the multiplier and new particles are not created as it becomes impossible for the creation to continue in this way. The creation stops abruptly and takes a different direction. In other parts of the universe, the creation continues to create the particles, but as it becomes clear, the creation cannot go past this point, and it stops completely creating particles and changes its direction to use the particles and create new entities.

The cessation of the creation of particles is not something the creation intended or planned. The nature of particles, the properties of the particles, and, more importantly, the events that followed, forced the creation to take a different path. The reason is the electrons: if electrons were not created or were created later, the creation had a chance to create other new particles, which would have resulted in a totally different path and a different outcome. On the other hand, if the electrons were created first ahead of anything else, the creation would have stopped and never created other new particles, but this could not have happened, as the electrons needed at least four particles in their composition and one has to be a failed proton or a proton particle. If the atoms, for example, had another subatomic particle in their composition, or a completely different nucleus was created, and they had something else in the core, and protons were doing what the electrons do now—orbiting the nucleus, it would have been a completely upside-down and back-to-front creation of something similar to atoms and molecules but totally different.

The electrons are the mongrel of the creation and bastard of particles. They never get along with each other and can never be in the same orbit with another electron. They need to be two on the first shell and eight in the next and last shells, and if the last shell is less than eight, then the atom has a problem and needs to share its electrons with another atom that has the numbers it needs to make eight. For example, Oxygen's atomic number is 8 (2, 6) and it needs two more electrons to complete its last shell (2, 8). Hydrogen's atomic number is 1, and it needs one more electron to make the two to become stable. One atom of Oxygen and two Hydrogen atoms share electrons to create a stable molecule that is known as H_2O or water. Sulphur (S) has the atomic number 16 (2, 8, 6) and if mixed with two hydrogen atoms at 200°C it makes Hydrogen-Sulphide H_2S, also known as Hydrosulphuric Acid.

Interesting fact: electrons were solely responsible for hurting 2.5m people in Great Britain in 2010 including 28 deaths, and the number of those who received serious injuries was 350,000. In the United States on August 6, 1890, the first person in the history was given a lethal dose of electrons by New York's Auburn Prison. You never hear that someone was "protocuted" because protons don't cause any problems! It is always the electrons that electrocute people.

We are about 1 trillion years into the creation of electrons and for the last time in the history of creation time ends. This last period of time lasted about 6 trillion years, looking at the long history of creation 6 trillion years seems very short for a universe. The reason for the stopping of time is there is a state of chaos in the universe, and the formerly steady movements are not steady and accountable anymore. The entities move erratically, and there are no signs of calm. The activities of the entities have been on the increase for the past 100 billion years but still time could have been recorded. The entities are moving from any direction and colliding into each other. Most of these collisions are not productive and only show a sign of chaos. The protons, quarks, and other particles are still created at the same time as the electrons. When the creation moves from creating one particle to creating a new one, it doesn't stop creating the previous creation, and they continue being created parallel and in conjunction with the new particles.

The creation continues to be productive and creates all particles on a large scale: all particles except Type-0. Although the Type-0 particles may well be still created by the Greater Universe but this universe cannot create them at all. The likelihood of the creation of the particles by the Greater Universe is very small, but it can restart at any time when the conditions are right.

The universe is becoming vibrant, and no sign of clouds can be seen. If there was a light source, it could be seen a long distance away. There are a large number of entities scattered all over the universe and every time they crash into each other, they create more heat, which increases the temperature level by a small margin. Friction is the reason for the rise.

As the result of the migration of protons, and later electrons, from one area to another, the quantity of mixed particles was on the rise for the past 3 to 4 trillion years. First, the protons spread all over the universe, and then the other particles and finally the electrons. On many occasions, extremely large objects—larger than the largest object we can identify now, made entirely out of protons are on the move for thousands or millions of years and smash into a similar size object made entirely out of electrons and create an even larger entity. These collisions cause the particles of various sizes and natures to be packed into a very tight and cold place. These collisions are not strong enough to create further events, and the objects just merge into each other. The creation of these super giant entities produces stronger gravity and causes the core of these entities to become more compressed. They also create a larger gravitational force in the area, which will clean up any loose particles in the space around. Almost no particles are seen wandering around although the collisions push a lot of particles and large chunks of entities into space. Eventually, these chunks and loose particles are joined to other entities.

A large amount of heavily compacted protons and electrons that could not have had any activities in cold temperature and high pressure and were still in a state of ZPE are freed from the entity by the energy created by friction that also causes a rise in temperature.

The energy created by friction releases some of the particles on the surface of the entities during the collisions and allows them to interact with each other. In very small areas of the colliding entities, with the rise of temperature, the protons and electrons interact and start a very short and small chain reaction which increases the temperature even further but fails to spread across the entities as the collision pushes them away and disbands the chain reaction. For a very short time, some of the electrons and protons react to create atoms and start a process that fails quickly and ends prematurely. This short event is the creation of the first nuclear chain reaction.

First Nuclear Chain Reaction

The first nuclear chain reaction created and released some elementary first atoms; these atoms are Hydrogen and Helium.

The completion of the first chain reaction takes thousands of years, and it releases some of the new atoms. It also burns off and destroys a lot of the particles that were created earlier. Temperature increases temporarily in a small area but goes back to cold later after the chain reaction ends. This chain reaction results in both Nuclear Fusion and Nuclear Fission. Nuclear fusion is a process when two or more nuclei are joined together to form a heavier nucleus. Nuclear fission is when the nucleus of an atom is split into smaller nuclei. This chain reaction contains both fusion and fission at the same time and in the same area of the colliding entities. In some moments, these two reactions, exactly the opposite of each other in the reaction chain, are caused by the impact on nuclei sitting side-by-side. This means that with two nuclei sitting next to each other, one is forced into fusion and the other one into fission at the same time. The parting nuclei are forged into the other one. The great force of the impact and small amount of nuclei taking part in this reaction stops the reaction becoming widespread, and it is forced to stop in a matter of seconds. These nuclei are the particles that were created earlier, and they were subjected to pressure, force, and compacted tightly together like any other particles we have now, except these particles have no nuclear and subatomic properties and activities. These are just like snowflakes that become hard and icy when compacted.

The sudden collision between the two extremely large entities pushes the particles even harder, and the heat created by friction is enough to raise the local temperature and allow the particles to become active; all this happens within a few seconds. Due to the energy created by friction, it would be probable for the particles to become active, but due to the other factors such as the presence of a single particle and lack of room for the particles to manoeuvre and pair with other particles, it is impossible for the event to complete.

Any collision has at least two phases and sometimes more. First, when the collision starts, two entities meet and come into physical contact, this phase continues until the energy that brought the two together is used. Although the entities may not come to a complete rest and continue moving in the same direction and merging into each other because the energy that brought these entities together and started the collision has ended so has the first phase of this event. To put it in perspective, it is very similar to firing a gun: the first phase is the gunpowder igniting and the bullet leaving the cartridge. The second phase is when the bullet arrives at its destination and starts the impact, and finally, the third phase is when the bullet comes to rest, and all the energy from the gunpowder is exhausted.

The entities act in a very similar manner to the bullet. The energy that makes them move comes from the gravity of the opposing entity, or one entity was moving and was deflected in a collision with another object, making it take a different path. The energies from the both sides bring the two entities together and start the impact. The energies are then converted into heat, as the result of friction, and the particles of the entities are brought to an active state. These particles, if placed together, would normally either be attracted to or repel each other. The force of impact stops the particles remaining as they were in a frozen state, while the heat and pressure are applied to them.

From this point, some of the energy from the moving entities is turned into heat, and some is stored in the opposite entity, and, in accordance with Newton's law, a reaction starts to the events. The particles in phase two of the collision start a chain reaction as soon as the pressure is reduced and the entities are moving in different directions. This chain reaction is very short-lived and contains both fusion and fission types. The reaction completely disappears as though nothing had happened when the entities absorb the particles. The second phase of this collision will take thousands of years to end completely. The energy released from the collision takes a long time to disappear because there is no other form the energy can convert to and it remains active until it is converted to a different form.

What makes this chain reaction so unique, apart from being the first time such an event happens, is the massive quantity of energy generated and converted to other forms as the result of this collision. There is a huge amount of energy stored into the two entities by the gravitational force pulling them towards each other. When the two entities meet at the collision point, the energy starts to be released and converted into heat. This collision is very slow and takes a long time to complete. The heat then causes the particles to become active and the nucleus to form. As a result, atoms are created. Simultaneously the energy is splitting the atoms and the nucleus apart causing a fission reaction and immediately fusing the same nucleus to other ones causing a fusion reaction.

If this event happens now in this universe, as we have experienced over the years, we would see a very large and destructive sudden release of energy larger than that seen in nuclear tests. Unlike a nuclear test, this chain reaction is completely contained and is isolated inside the colliding area of the two entities and, due to the large energy pushing these two together, the energy from the chain reaction is dwarfed and trapped by the colliding entities and cannot be released. If we could test this in a lab, splitting the nucleus very slowly and in a controlled environment, and could control the energy released from the splitting atom; we could solve the world's energy problem. However, the release of energy is so sudden and powerful it would be impossible for us to contain it.

Would it be possible to build a small device, small compared to the nuclear power plants, to split one single nucleus in a fission process, force it into another nucleus in a fusion process immediately, then harvest the released energy, and continue doing it again in the next cycle a minute later? A continuous fission to fusion nuclear process which is not a chain reaction due to the absence of nucleus numbers but will use the same nuclei over and over again. This is definitely not possible as the energy released from the fission is used in fusion, and will not sustain a long process. We use a bombardment-driven process that results from the collision of two subatomic particles, and this makes it impossible to build a small device, which is not a reactor and is more of an energy cell. In the nuclear reactors catalysts like Boric acid (H_3BO_3) is used to control the speed of fission, but at this point in creation there are no catalysts and the creation just tears apart the nuclei and pushes them into each other by sheer force.

Cold Fusion is another method of splitting an atom. It is a very problematic process and would occur at room temperature; as opposed to the hot fusion that happens within stars. The process involves electrolysis of heavy water on the surface of a Palladium (Pd) rod and releases hydrogen from heavy water.

Cold fusion power units are already built and ready for dispatch. E-Cat (or Energy Catalyser) is a cold fusion power station built by Leonardo Corporation in the US. Leonardo Corp. is the company behind a 1-megawatt power station built into a 20-foot shipping container, and the cost of electricity produced by this device is $1 per megawatt hour, compared to $100 per megawatt hour for coal power. This company has also built a small 10 kilowatt unit for home users the size of a bar fridge. Although these units seem very attractive in upfront investment cost and the running cost as compared with any other source of energy, it seems very unlikely that governments will allow the sale of such devices due to security concerns. The unit itself is tested and is very reliable for the home user and can be used as a small but effective tool, but due to security concerns and lack of security on site, could create panic in the event of a terrorist attack. The news of leakage of nuclear fuel in suburban homes would have a greater impact on the public than the attack itself. Also, the sale of the larger 1-megawatt units to large industrial manufacturers or corporates is a greater security concern.

Although this chain of nuclear reaction took only a few days and ended quickly, it was enough to release a large amount of heat, which quickly travelled through the entities and increased their temperature considerably. The heat was enough to activate some particles. These large entities are now composed of all the types of particles created so far. The particles are sitting side-by-side, compacted, under extreme pressure and low temperature and have no chance of any activity. During the collision, these particles were exposed to higher pressure and higher temperature in which, as soon as the collision eased, the pressure was reduced, but the temperature remained high. This event allowed the particles to gain properties, become active and pair-up with other particles to create atoms. These atoms, very basic at first, were the very first atoms created and were all Hydrogen and Helium. The newly created atoms exist in all three forms of gas, liquid and solid state in the same area of the universe, and this is due to the heat having a scattered distribution. Many of the new atoms start becoming extremely hot and creating a glow, but they can not burn as no oxygen has yet been created.

The heat continues to have an increasing effect, and the entities continue to collide more often. The entities at this time of creation are almost entirely made of protons, electrons, neutrons, and other subatomic particles. Every collision results in the release of large amounts of gaseous and solid atoms. The solid atoms soon turn into gas as the temperature rises. The heat is warming up the universe. The hot gases are now fully active and have created light. The light is continuous and will not end. With the accelerated rate of creation, the heat is increasing, and the light is becoming brighter.

At this point, due to continuous subatomic activities, Time starts, and this time it will be permanent, time will not end from this point onwards. This is the time that can be counted as the start of the new creation and the time, which led to the Big Bang. The Big Bang was just a climax of the event started at this point. This is the point the creation of the new universe starts. The entities we had until now were composed of particles that were never forged together, they were just pressed hard into a large object, and not holding together with any glue, any hit could break them apart. From this point in creation, the subatomic activities are permanent, and the temperature has risen. This allows the particles, frozen before, to come to life and create new particles called Molecules.

The continuous light starts when one of the large entities becomes wholly active from the heat and the subatomic activities. The heat covers the whole surface of the object and makes it glow very brightly. This is the birth of the first star. The heat on the surface of the new star slowly penetrates deep, and the particles become active subatomic particles and start a new chain reaction. This chain reaction takes thousands of years to engulf the new star and eventually ends its life prematurely by causing it to explode. The explosion pushes the star's broken pieces into other entities nearby and causes those to become active stars. While the new stars continue to light up the universe and warm the surroundings, other stars start to be created.

Slowly, over the next billion years, part of the universe starts to have numerous stars. These stars collectively light up the universe and the light beams by this time have travelled more than a billion light-years across the universe and with it the packets of energy or photons have carried the heat.

The heat, which started on the surface of the objects, quickly engulfs the complete area of the object and turns it into a fully active object similar to a star. Over the tens of thousands of years, the heat slowly penetrates deeper as the particles below the surface come out of their frozen condition and become active. The object, after more than a hundred years, is completely transformed into a star and the below surface activities reach to the core of the new star, the core at this time is extremely dense and fluid. The lava core could not explode in a similar way that had happened in other stars when the chain reaction reached the core and caused the breakup of the star. Instead, the lava core pushes a continuous and strong internal pressure towards the surface of the star and causes the new star's shape to change and become round. The pressure comes from the molecules expanding; in order to bind the atoms and complex molecules, they need to move away from their position, so the electrons can freely spin around the protons, bind with other atoms, and create molecules. The pressure pushes the super-heated and molten object outwards, so the star becomes a sphere. This is not quite a new sphere, because the protons and electrons were created earlier, and they are spherical, but this is the first large object in this shape.

At this time of creation, molecules are created, and the further interaction between them leads to the creation of complex molecules. With the creation of complex molecules comes heat resistance. The entities enter a new phase as the result of the changes in their composition; the new molecules are resistant to temperature and show no signs of weakness. As a result, activities slow down, and the temperature drops, the new star stabilises and continues to emit a large amount of heat and light, brightening the universe. The sheer size of the star is beyond anything ever known, but still, these giant stars gain stronger gravity due to the molten core, and this causes the gravitational force to reach even further distances. With the wider and stronger gravity, the stars pull all the entities around and continue to get even larger.

The molten and almost fluid stars at this time start having magnetic force. These stars, though, are as bright and hot as the stars we have now in our system, but they are not quite like them. Our sun is made of almost 75% Helium and 25% Hydrogen and a small percentage of various metals. The new stars at this time of creation have a larger amount of metal and other debris on the surface due to the constant collisions of other entities; this causes a thick layer on the outer surface of the large entity and reduces the temperature. The metal surface acts as a protective layer and slows down the fusion underneath it.

Complex molecules are the very last things created: from this point on nothing else will be created. Creation stops completely, and whatever is created until now will shape or mix with other particles of entities to form and create new entities.

The new creation is not a real creation but the continuation of what has been created until now. The new creations are based on, and dependent on, the old created elements and entities, no new particles or atoms will be created for the new entities.

The end of creation is due to two factors: the first is a shortage of particles, whatever was in the universe is used and transformed into new particles and subatomic particles; and the second factor is that the temperature has risen. With the rise of temperature, it has become impossible to convert the Type-0 particles to subatomic particles, the heat increases the activities of the other particles and disrupts the process of creation.

Although the creation of particles has come to a complete end, not all the complex molecules have been created yet. The complex molecules are not part of the creation; they are the third phase of creation and are not a real creation. They are just a mixture of random atoms forced by extreme pressure as the result of collisions and gravity into a new combination. The majority of the creation of complex molecules is after the Big Bang event. The atoms are the second phase of creation, and they are fundamental to any molecular structure and any entity after that. The first phase was the Type-0 particles up to the creation of large entities, and right up to the creation of quarks and Dark Matter.

With the creation of stars and molecules, magnetic energy comes to life. Magnetic energy is the result of a magnetic field, which is created by electrons. This field has a North – South direction and the field creates energy within the field, which is also called potential magnetic energy.

From this point, the entities continue to collide and merge with larger ones, and stars start pulling smaller stars and consuming them. Some of the collisions are coordinated between a large number of stars, and the collisions are similar to a mini big bang. Many small and short duration Big Bangs happen during this period. This is not the only time mini Big Bangs occurred, in earlier stages also there were some minor Big Bangs, in the form of large to extreme collisions. This process of collisions and minor Big Bangs continues intermittently and will last for at least 2 trillion years. During this period, the stars become greater and greater by consuming other entities.

The size of these stars is staggering, even by the universe's standards. The universe has never had such large entities before. Some of these large stars may well still be present in a very remote area of the universe where the expansion of the Big Bang has yet to reach. Due to the distance from our planet, we are not likely to receive any light beams from them for billions of years in the future. We can fast-forward this period and have a look at the other side of these long 2 trillion years of collisions.

This long period of collisions has produced even greater stars and extreme entities. The universe is completely burning. Large magnitudes of light brighten up the universe, and temperature has risen to thousands of degrees above zero in large parts of the universe. Hardly any entities are left alone, the extremely large gravitational force is pulling everything, and stars continue to consume each other. Extremely large stars are formed, and they continue to pull other stars.

At this time, for the first time, a solar system is formed. This is the very first solar system in the universe and possibly the very last one too. This kind of solar system is unique; all the members of this system are stars and none of them planets. The number of members is not known, but whatever it is, it is a solar system. This system, like any other systems, has a giant star in the centre and many entities are orbiting it.

Although the distance between each member and the star in the centre is extremely great, it is the extreme gravitational energy of both entities—the star in the centre and the system members, which holds this system together.

The sheer gravitational energy and the centrifugal force are imposed at the same time on the members of this system, and it helps the entities to remain in an orbital path. The orbits of some of these entities are extremely wide; for some of the outer members of this system, it takes tens, to hundreds of thousands of years to complete one full circle.

The time when the first member of the first solar system completes its orbit and is back where it started its journey around the star, is the creation of the first calendar year. Time started trillions of years earlier and now time can be observed by calculating the number of seconds it takes for a member of the first solar system to complete one orbit.

This solar system is disrupted after a few billions of years by a large star that, travelling from far away, collides with the stars nearby and causes them to crash into this system and completely disrupt its operation. Some of the orbiting stars are pulled towards the centre of the system and those on the far side of the system crash into other systems nearby, causing a chain of collisions that lasts millions of years. With the collapse of the first solar system, also the first calendar year ends. The collisions with the large star at the centre of the system first cause part of the star to be broken and pushed far away, and then the gravitational force brings them back, and the star re-forms with new collided stars added to it making it the largest star.

Because of the collapse of such a large solar system and the collisions following, a black hole is formed. This is the first black hole created. However, it will not be too long before this one also collapses. Black holes normally last a very long time due to their nature and the fact they are constantly refuelled by consuming systems and galaxies. Not this time; this black hole is too close to events taking place, and the collision this time swallows a black hole instead of the other way around. The black hole does not last more than a few hundred thousand years and completely disappears when it is consumed by the mega star.

From here, events are happening very fast, and whatever is created does not last long and is completely transformed to something new or consumed by something bigger. The extreme gravity was created some time ago, and it is pulling many entities toward each other. The end of the creation of particles was earlier, but the particles continued to keep the process active and press ahead with the events. The end of this stage is very fast approaching, the events happening at shorter intervals. These events are small and short compared to what will come next. At this period of time, a larger number of extreme mega stars form. These stars are many times greater than the stars created earlier. The large stars, created before, generated an extreme gravitational force that pulled them together and caused a collision that resulted in entities binding and creating the extreme mega stars. With the creation of new stars, the gravitational force also increased, and with it came more pulling power to bring more entities closer.

The sheer size of these extreme mega entities is staggering, but it is not going to work favourably for them. Every time one of these extreme mega stars is created, its gravitation pulls in more entities. Some smaller ones just crash into the entity, but others are just too great and powerful to crash and disintegrate; they cause the complete destruction of both objects and create a very large destructive wave and explosion. Not only does it destroy both, but also the debris continues to destroy the neighbouring objects and crash them into each other. A large part of the universe is covered by large pieces of former stars burning openly. This burning is not like an open fire with flames, smoke, and ashes; it is an open nuclear chain reaction continuously reacting to the pressure and sudden release causing the molten objects to scatter like lava and evaporate like ignited gas.

A red-hot fluid-like substance covers a vast area of the universe, and it is in a state of nuclear chain reaction for quite some time now. Entities travel relatively short distances from nearby and, when they arrive in this area, they just disintegrate and add to the existing molten substance. The creation, until now, was bringing the particles together, combining them, and creating larger entities; but what is happening now is the opposite of what happened before. The gravitational force is bringing the large entities together, and the high temperature and the open chain reaction is causing them to break apart and join the substance. This is, in part, the opposite of what has been happening until recently.

There is chaos in the universe. We are at the last part of this tier, and the Big Bang is approaching. The stage is set, and the event has begun. From galaxies, billions of light-years away, extreme mega stars are on their way to meet at almost the same point, and with them come large contingents of smaller entities. The event started not long ago, and the extreme gravity started pulling objects, and these objects joined other objects to create even greater objects, which caused huge collisions. At the same time, there are entities broken up from earlier great collisions heading in all directions and with them, very large amounts of energy are transported. The event of the Big Bang is not something that happens overnight and completes by morning. This is going to take millions of years leading to the big event and millions of years after the Big Bang. At the beginning, the collisions start small and a smaller amount of energy is released from each collision, but it gets to a point where the colliding objects are extremely large and have travelled a greater distance. This adds to the amount of energy that is stored in them, and at the time of the collision, more energy than usual is released.

The entities pulled towards the area of the universe where the open nuclear reaction is in progress, are a very small part of the entities that will be involved in the process. Most of the entities not involved with this will later become involved as the particles from the event will reach them and will create a chain reaction. It is not as though all the entities from all around the universe come to the centre and erupt. The event starts in one place as the result of collisions, larger than were happening before, and the particles from this event reach other entities, and the chain of collisions continues to engulf the universe with an open nuclear reaction such as the universe has never seen.

What triggers this event, besides the extraordinary size of the entities involved, is the composition of the new entities that came from far away. These entities carried new molecules and atoms, not friendly to the existing entities. The mismatch in their composition created a massive chain reaction, very quickly, the reaction reached the other entities nearby, and suddenly there was an elevated amount of reaction. The entities arrived in the area that was already in a state of nuclear chain reaction. The newly arrived entities were not active at all, they had a different composition to the existing entities, and the sudden introduction of non-active entities to a very large field of openly active entities caused a very large disturbance.

The sudden release of a massive amount of hydrogen and the sudden entry to the open nuclear chain reaction releases an extraordinarily large amount of energy.

The event of the Big Bang is not the whole universe blowing up like a giant explosion. If it were, it would leave an even larger empty space at the centre of the explosion. Although there is a rough centre of the universe, indicating it started at this point, and the entities are moving away at a similar rate, and in all directions from a similar point, this point is not the one point that blew up. The ground zero of the event is not one particular area; it is a very vast part of space occupied by a concentration of mega super-giant entities all in a state of nuclear chain reaction. The explosion event took millions of years to reach all areas from this concentrated active area.

This event has been in the making for the past billions of years to allow the entities to travel from the furthest parts of the universe. The time for the actual event of these large collisions, some call it the Big Bang, takes at least 100 million years to complete. The event is not something we can mark on our calendar as one event. The event is a combination of a series of events, from the changes leading to the event billions of years earlier, and the progress of the event from start to completion of the event; and finally, the aftermath of the event hundreds of millions of years later. In total, it will be at least 4–5 billion years. We had seen the end of it when the stars cooled down and turned into planets, and the planets cooled down and finally life began on some. In one sense the event has not completely ended because remnants of the great explosion are still visible, in another sense, what we see today may not be the result of the big event, the universe started long before the big bang and the big event was just a small bump on the road to eternity.

This is it, we are at the end of our journey at the main event, and the fireworks have just begun. It took trillions of years and a very slow process for the creation to come this far. The result compared to the effort, time, and energy put into in the creation, may well seem not worth the efforts from a human point of view. The creation ended more than a trillion years ago when the molecules were created; and this, the big event, is the beginning of the end of the creation. What has been achieved to this point, it is going to blow up in a few days, but some may say the creation is the transformation of itself and the creation of life is the proof of that.

If you look at the whole process as one event, you may see this big event as the peak of the creation. Actually, the peak was when the complex molecules were created: it was flat-lining from then, and it will go down all the way to the end. We will see some transformation of the molecules and some new ones created after the big event, but the structure of the new creation is based on the same old creation of the old atoms and molecules, even the new molecules are created from the same old atoms. No really new particles are going to be created, but the transformation of the creation is staggering and mind blowing. Nobody thought that life could be created from those little particles.

Next Event: Creation of Life

CHAPTER 10:
HOW WILL IT END?

The Big Bang was the beginning of the end for this universe. The creation passed through two stages and four tiers, it has reached its apex and conquered its climax. It created all the particles, and matter from particles, and life from matter. In one sense, the creation is the cause of its own death. The particles were created for some reason. They sat there in a cold, dark, and empty space for trillions of years until they started moving. One thing led to another, things were created, and then, finally, the whole universe blew up. The Big Event started some fires, pushed the stars away from each other, collisions reduced, and space started to cool down a little bit. Stars burnt out and died. Planets were created. The planets were smouldering at first, and then they finally cooled down. Life started on some of them, or at least on one we know of, and they, also, finally died when the star in the centre of their system died. This short detail concludes the 20 billion years' history of the universe from Big Bang until the death of our solar system. It is hardly explaining the event thoroughly. Life in other solar systems may continue for billions of years more, but those also will end. The life created by this universe is not meant to be forever because the universe itself cannot be around forever. Life of any kind has an expiration date and a period of living. Sometimes, this period is cut short by other factors and sudden events beyond the capacity of such life. The word "LIFE" is very widely used here and can refer to any primitive.

It will never end for the Greater Universe, but it will transform into another form and will continue to exist. The existence will never end. It will be a very long ending to the current life of this universe. The universe will transform itself into something new. It may end for humans and life of any kind but not for the universe. It will take perhaps hundreds or thousands of Cotillion (10^{303}) years to complete, although all the molecules in the universe, which are dispersed protons and electrons, will not be destroyed completely, even the decay of particles will leave decayed particles behind. Although it is said, the protons have a half-life, a lot of energy is required to break the protons and electrons. At this stage, no energy is left to cause a complete dismantling of the universe. By this time, a new universe starts to form at the other side of the Greater Universe. Remember, earlier, something created the particles thousands of trillions of years ago. Most likely it still continues to create the particles, and there will be a new process of creation, only this time a lot of electrons and protons are already in the system, and the creation will take a different path and will create new molecules.

Stars burn out, planets crash into each other, there will be no lights, and energy will end without a source. There will be a sort of zero-energy environment after Decillions (10^{33}) of years. Some of the subatomic energy will remain active, as there will be no pressure or a high enough temperature to stop the activities. Some protons and electrons will be separated by a flood of particles but eventually a small number of them will bind. Energy will not completely disappear; it will remain at its lowest form in a very small scale just between molecules.

Japanese scientists in 1982–1983 conducted an experiment called "Super-Kamioka" or in its full name: **Super-Kamioka Neutrino Detection Experiment**, the aim of this experiment was to search for proton decay. This experiment showed that a proton has a mean lifetime of 6.6×10^{33} years for decay. This means the proton will be neutralised and lose energy after many years but will not be completely banished. The protons may decay, but the remains will not turn into the Type-0 particles. As discussed earlier, the protons have properties, and the particles don't; the properties of the broken and decayed protons remain. In a similar way, the broken protons in a CERN experiment continue to adopt the properties from the particle; the decayed protons also continue to have their properties. The properties of the entities we have now on our planet are identified by the molecular structure of the entity; the same molecule that makes up the entity exists in smaller pieces of the entity. The protons and electrons also have a structural composition in a way similar to a molecular structure and they, also, can be identified from the broken pieces. Because we don't have the ability to determine those structures it doesn't mean they don't have them. As discussed earlier, the subatomic particles are composed of smaller particles of Type-0, quark, and Dark Matter; and the multiplier number of the mixture is what defines and differentiates them. With the broken pieces of protons or decayed particles, if we could devise a suitable test, it would be possible to test and to determine the composition of the decayed particle. It would be interesting to see what a large amount of decayed protons looks and feels like; it may feel like the finest powder of any kind in the universe.

The end, for this universe, is painfully long. It is longer than the time it took to create it. It will take almost forever to complete this task if it is ever completed. The cold will not come back to this former universe. The temperature will remain at absolute zero, or just above, but will not go any lower. The loss of energy by entities and subatomic particles is enough to bring the temperature down to this point but not any lower. Every now and then, a particle or a larger part of a former entity will continue to pass by the remains of dead planets, but even those will eventually cease to exist and stop by running into something.

Last Words

On the subject of gravity: a force exists between the atoms and the molecules besides the one that binds and holds them together. There is an energy, extremely small in size, which cannot be measured because of its size, and it cannot have any effect on other subatomic particles or any object we see. Collectively these small energies work incrementally and can create a stronger attraction force we call gravity. Interestingly, this energy has no negative/positive or south/north rules: it just attracts to a similar neighbouring energy. This energy, in a cumulative form, has a natural centre and the attraction always pulls towards the centre of this field. This energy is so small that, even collectively, it remains small.

If we have two bricks, for example, put together in space where there is no influence of gravitational force of any kind, these two objects will have absolutely no effect on each other; at least, none that we can measure. If we increase the number of one of the two objects to 1000 trillion of bricks, then this large mass of bricks will have a gravitational force, which comes from the collective amount of the same energy in every individual object. This gravitational force increases further if we increase the density of the bricks by compacting them more. We can measure this gravitational force, divide it by the number of bricks, and have the amount of individual attraction energy of a single brick and thus a single atom.

By the end of this century, the future generations will be able to build an instrument sensitive enough to measure the rate of Subatomic Attraction Energy (SAE) between two objects and it is hoped they can build a device to have a counter-effect and cancel the energy. This device would be similar to noise-cancelling headphones. These generate a signal of the opposite phase on the same frequency as the signal they are cancelling. Similarly, we could build a device to create an opposing energy or wavelengths of the energy or wavelengths molecules and atoms generate to create gravity, and so cancel their gravity. Of course, the device we build is going to be made from the same atoms and molecules that create the same energy in the first place. If we could, perhaps we would create new molecules and atoms and use those to make a device or create an energy enough to cancel both energies of the device and the objects to cancel the gravity.

First, we need to know the amount and characteristics of this energy, all we know until now is something is pulling us down, and we don't know its source, we just call it gravity. Although gravity can fluctuate on the same or different planets based on the density of the entity, the rate of SAE is always exactly the same on a subatomic level. The SAE exists across the universe in every atom, and its amount is universally equal for the atoms of a particular element. Atoms with a larger and heavier nucleus have slightly larger SAE than smaller atoms with a smaller nucleus. Hydrogen atoms with no neutron and Helium with two neutrons, for example, have the lowest rate of SAE whereas a Uranium U238 atom which has 146 neutrons has a higher rate.

The extent of attraction energy of one atom here on Earth is exactly the same as in an atom on an entity right across the universe on the other side 13b light-years away. We need to know the rules of this energy: whether it has frequencies or; similar to a magnetic field, has south/north poles; whether it is is negative/positive; if it has directions; and any other information we can gather. Perhaps the future generations can use the Proton Milling Machine and make some proton powder, make some new atoms and molecules from it, and make a device from the powder to build an antigravity device to cancel out the gravitational force. The device might also work in reverse, act as a source of gravity, and create an artificial gravitational force where it doesn't exist, or exists but is weak, such as onboard spaceships on long-haul interstellar travels. The key to this method is to find the source of the energy at the subatomic level.

The gravitational energy comes from the core of molecules and atoms. The denser the object is, the stronger the gravitational force is. This demonstrates that the closer the nuclei are to each other the more the combination and collective attraction energy can increase, and the further apart the nuclei are the weaker collective attraction energy and decreased gravitational force is. Saturn, for example, has a larger size than Earth but has lower gravitational energy due to Saturn's lower density. We need to determine the amount of energy from a single atom and find out if the energy is completely from the nucleus or from the whole atom. It is not from the molecules as the molecules are a combination of atoms. If the energy comes from the atom, then it has pulse and frequency; and if it comes from the nucleus, it has no pulse. In most tests conducted in labs, hydrogen is used to separate the proton from the electron because hydrogen is the only element without a nucleus: at the core of the hydrogen atom is only one proton. It would be interesting to know if a large amount of solid helium and solid hydrogen would show signs of gravitational energy as opposed to uranium with the same number of atoms. This test would show if the atom without neutrons and only one proton, or the atom with two neutrons, or the atom with 146 neutrons had the higher rate of SAE.

The strongest possibility is that the energy comes from the nucleus and has no pulse. If it did come from the atom and had a pulse, the pulses could influence and cancel each other out, or increase and boost the gravitational force. In this case, the gravitational force could fluctuate, and the changes in entities' structures could have a large influence on the amount of gravity. As we experience daily, gravity is always exactly the same amount everywhere with a similar density, and there are no fluctuations in different places on the planet. Having said this, we should also say the gravity had existed trillions of years before even any subatomic particles were created. The Type-0 particles did have an attraction power when they were in a greater density and became extremely large entities. It is very much possible the SAE comes from all particles in atoms due to the composition of particles and the fact they all have Type-0 particles in their structures.

Defeating gravity and having control over it will help humanity in many ways. It should also be acknowledged that building an instrument sensitive enough to detect this energy at the subatomic level, is a complicated job and, with the current technology, it remains a distant wishful thought.

Problems with Understanding Gravity

This illustration, the original artwork of NASA, is to illustrate how the gravity of planets dimples the space around them as Einstein said in his Theory of Relativity, like a ball on a trampoline and causes the light to bend when passing by the planets or black holes.

Figure 40: Spacetime Dimpled Around the Earth

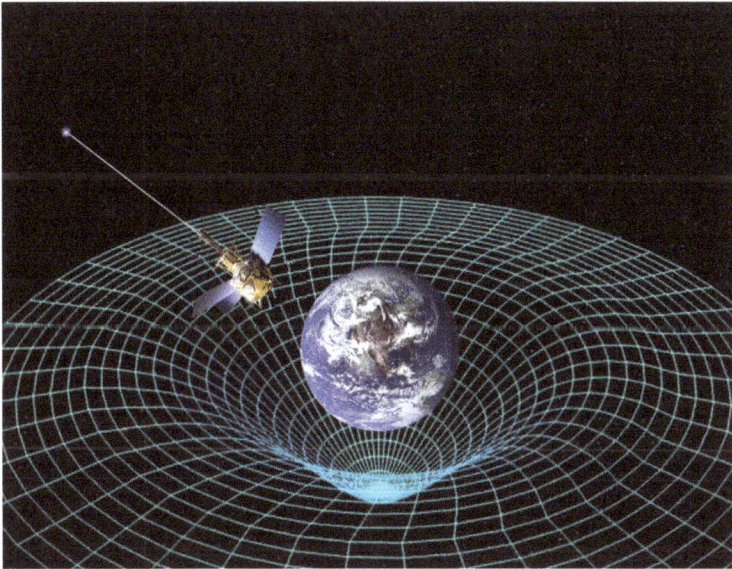

As he worked out the equations for his general theory of relativity, Einstein realised that massive objects caused a distortion in space-time. Imagine setting a large body in the centre of a trampoline. The body would press down into the fabric, causing it to dimple. (From: space.com)

NASA experiment on light bends due to gravitational force of the Earth and the vortex it creates:

May 4, 2011: Einstein was right again. There *is* a space-time vortex around Earth, and its shape precisely matches the predictions of Einstein's theory of gravity. (From: http://science.nasa.gov/science-news/science-at-nasa/2011/04may_epic/)

Figure 41: Incorrect Representation of Spacetime Being Bent by the Earth

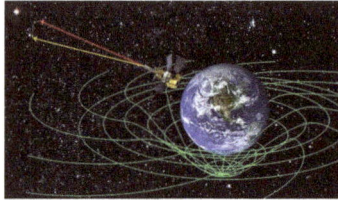

These two images, and many more in the public arena (some are shown below), wrongly illustrate the effect of gravity around the object in space. It is not necessary to discuss the push or pull effect, as you are all familiar with it; but all these images, texts, and literature show a push by the planet or an object instead of a pull effect. A push by the planet's weight rather than a pull by planet's gravity. Planets do not have any weight in space other than their gravitational attraction to the Sun, which is offset by the centrifugal force, but they have mass. Nothing has weight in space, but the planets have a lot of mass. You place the same ball on a trampoline in space, away from the earth's gravitational field, and you will see no dimples, whatsoever! The ball will hover right above the fabric where it was placed, and will not move unless it is moved using external force.

If gravity has any effect on space around it, and this is a totally different discussion from the theory itself, shouldn't the illustration at least look like the picture below? The dimples are towards the planet, not the opposite direction and away from it:

Figure 42: Gravity Pulling Space-Time Towards the Earth

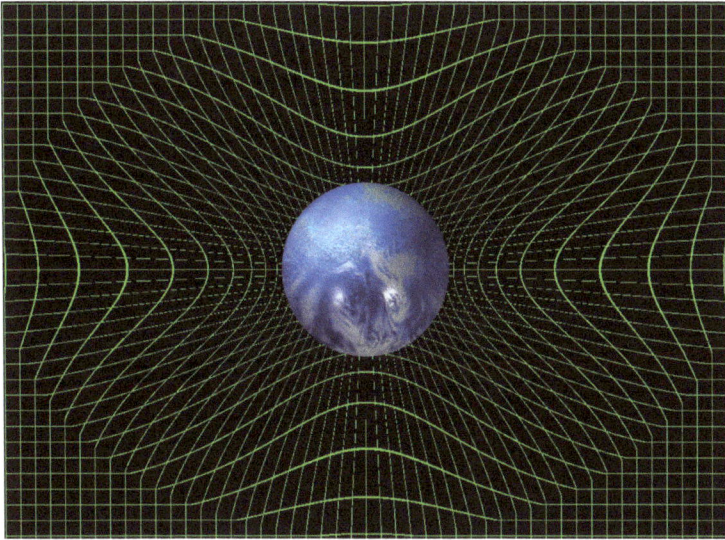

The gravity would pull not push the space around objects, and all illustrations so far point to a "push" effect. The ball on the trampoline is dimpling the fabric because the source of gravity is outside the ball, in this case, the Earth, and correctly said it dimples the fabric. When the source of gravity is not outside the object, and instead the object itself is the source of gravity, the fabric should not only not dimple, but should wrap around the ball in equal proportion to the force of the pull. Below are some more illustrations, found in the public domain, which show the deep misunderstanding of the theory. This mistake has also been made by the owner of the theory, Einstein himself, or people misunderstood the concept.

Einstein in his Theory of Gravity specifically talks about gravity and not mass. If the ball is placed over the fabric in zero gravity, both remain unaffected by each other's presence, but, if the ball is moved towards the fabric by an external force, then the mass of the fabric will resist, as the mass continues to resist any changes to its state of rest and this will cause a dimple. This dimple is not caused by the gravity of the ball, but by the external force applied to the ball and the resistance of the fabric's mass. This is called inertia, and inertia is the resistance of any physical mass to any changes to its state of rest. In the theory of gravity, the planet's dimple in the space around it is specifically related to the gravity and the weight of the planet, not the mass of it.

In these illustrations, the space around the planets should be pulling up towards the planets not pushed down by planets. The grids could be placed above or below the planets in the illustrations, but the dimple or deflection of the grid will always be towards the planet.

These illustrations display the exact opposite of the actual effect of gravity:

Figure 43: Incorrect Illustrations of the Effect of Gravity

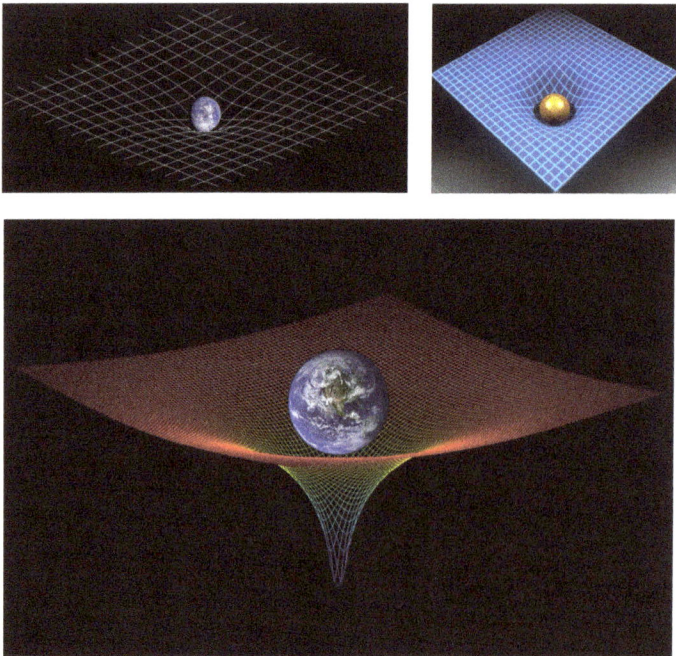

If the Earth caused the dimple, influenced by the gravity force of another object, like the Sun, then it would be right to illustrate it in this way. However, the effect of gravity from the Sun on the grid would be equal at all points the same distance from the Sun. The dimples shown in these illustrations are placed on the wrong side of the planet; they should be placed at the top of the planet not below. The dimple should be shown all around the planet not only on one side. The gravitational force is applied in all directions and in an all-dimensional environment. A reminder: there is no up or down, left or right in space; everything is in the same space, and it depends on the direction of the observer looking at it.

Now to the dimple itself: the dimple, and the ball on the trampoline, as Einstein described it, is a visual aid to help understand the effects of gravity. Space, as discussed, is not a physical entity and cannot be bent, dimpled, tunnelled, or folded. The dimple comes from the invisible force-field of the gravitational force, and the force, similar to the electromagnetic force, creates directional and invisible grooves in a field in space around the entity. The electromagnetic field can be mapped using a compass, and, on a small scale, a magnetic block can be used to attract metal sawdust to demonstrate the field. Sensitive instruments can measure the magnetic flux.

Figure 44: More Incorrect Illustrations of Gravity

The gravitational force always pulls an object, and if there is some object in the middle that causes the pull to stop, it will create a dimple if that object is soft enough to bend. The planet's gravity cannot bend space like a trampoline as it cannot pull itself like a third party, but in hindsight, it can pull the space around it; and when it pulls, the dimple doesn't look like the ones widely available in the public domain. The trampoline is also acting as a resisting device against the gravity force pulling the ball, and, because of the resistance, the trampoline is dimpled. In space, there is no resisting device and the space around the entity having no anchor points to hold on to, resists the gravitational force to pull the entity. In space, there is no resisting force, and if anything is moving, it can theoretically move forever. There is no anti-gravitational force in space against the planet, and there is no need to illustrate a dimple in such way. The way this theory is illustrated in the public domain gives the impression the whole universe is placed on a flat fabric-like environment, and an invisible force of unknown origin pulls everything that moves. If the weight of the planet is pushing down the space around it, and causing the dimple, then where is the source of the resisting force pushing the space up? We know the source of the resisting trampoline stopping the ball; it is the legs of the trampoline holding the fabric and allowing an anchorage for the fabric. Space is not matter, or an entity; space physically doesn't exist, it cannot be bent or dimpled or pierced, and this should diminish the last hope for the wormholes (also there are no parallel universes).

Interestingly enough, the planets, stars, or any other entity in space, including the Earth, has no weight whatsoever. If a person can be weightless in space so can be planets, there is a reason; it is called "zero gravity". The weight is defined by an object on the planet and the planet acting as a third party and a source of the pulling force or gravitational force. The entity itself, on its own, cannot create a weight for itself, as it cannot act as an object and a third-party source of gravity at the same time. The planet can be subjected to a gravitational force by another entity and have weight, but not on its own alone. A planet far enough not to be affected by another body's gravitational force has no pulling power over itself and cannot create its own weight. It is like a person pulling himself across the room using his own arm, or using a scale to weigh itself. No entity can create its own weight; the weight is created and influenced only by an external force. The concept of the weight of a planet dimpling the space in such a way as illustrated above is just incorrect.

The gravitational force pulling-field around entities is not physical and cannot bend light, as the photons have no mass, and cannot be affected by gravity. However, the gravitational force pulling-field creates a groove-like force-field, allowing light to enter the groves and be directed in the canals the grooves create, similar to a driveway pavement where water runs between the brick joints. These gravitational force pulling-fields can be anything between hundreds of thousands of kilometres to hundreds of millions of kilometres wide depending on the size of the entity. The gravitational force pulling-field slowly disappears and turns to zero the further it spreads. The gravitational force pulling-field is wrapped around the entity similar to the atmosphere except it reaches further into space and shapes like a sphere as the entity is.

Sometimes these grooves act like a prism and change the wavelength of the light beam causing a redshift without fracturing it. A prism fractures light, but grooves cannot fracture light. On the outer edge of the gravitational force pulling-field, where the grooves become weak, the light crosses and jumps these grooves. The grooves act like a corrugated field and cause light to change its speed and distort its wavelength. The inbound light beams are also affected by the corrugated grooves when they enter the gravitational force pulling-field and they enter with their wavelengths changed and when they go out from the field it can magnify the redshift or decrease it depending on the angle of the grooves.

Once again, we are presented a theory with a wrong illustration. This picture below was presented by NASA on Feb 12, 2016, announcing ripples in the fabric of space-time known as gravitational waves first proposed by Albert Einstein 100 years ago:

Figure 45: Gravitational Waves

The ripples in the fabric of space-time were discussed earlier, but this picture is fascinating to see and looked very familiar.

The illustrations below show the magnetic fields around the planet Earth and the Sun's magnetic effect on those fields.

Figure 46: Magnetic Fields Around the Earth

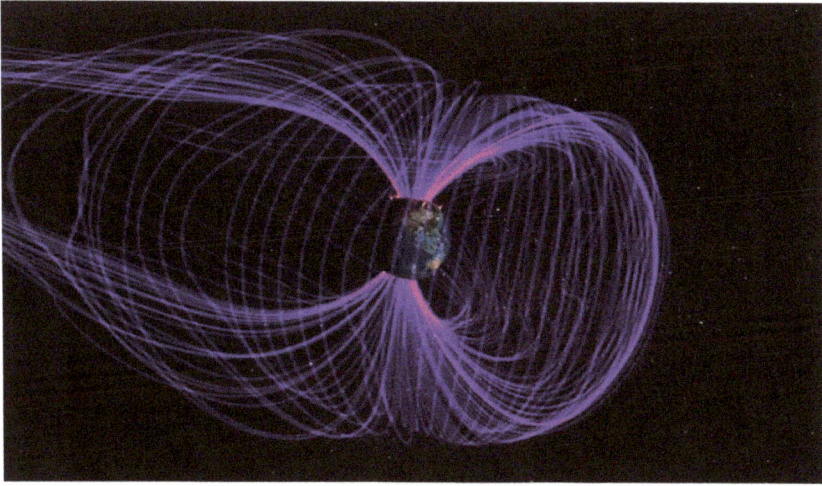

Image Courtesy of NASA

The illustration released by NASA in Feb 2016 is more like the electromagnetic field than a gravitational field. Gravitational fields, unlike the electromagnetic fields, do not have multiple origins on the same entity; the whole entity is the origin of the gravitational force field. The illustrations presented by NASA show multiple start and end origins from the same black hole. One black hole has only one sphere-like gravitational field. This field generates a very strong pulling force from all directions towards the centre of the object.

NASA and ESA used laser beams to capture the gravitational ripples and show that space-time exists. The method they used is very accurate, and the instruments are state-of-the-art technology. They observed the collision of two black holes 1.3 billion light years from Earth and, by using a laser beam, they were able to capture a difference in time due to the effect of the gravitational wave on the time taken by the laser beam to travel A to B. The test showed a disparity in time for the same amount of space travelled by the beam of light.

The distance between A and B had not changed but the time for the light to travel this distance was a very small amount longer. It took the light longer to travel the same distance. This, as they say, is proof the strong gravitational force can slow light, and thus slow time; and that this shows the gravitational ripples exist as Einstein predicted 100 years ago.

Figure 47: Sun's Effect on Magnetic Field of Earth

Source: physics.unlv.edu

Well ... as discussed earlier the gravitational force pulls from all directions towards the centre of the entity and these forces are fields of energy. When two or more fields of energy collide, the direction of the energy changes at the point of collision, and it goes back to its original direction once it has passed that point. The gravitational fields, similar to electromagnetic fields and other energies on Earth, travel through the air like sound and shock waves. They do create a ripple when they overlap each other, but the energies created by this overlapping and the collisions of black holes are directed inwards and absorbed by the black hole and will not travel outside the event.

The picture below shows the collision between two black holes and the maximum distance the gravitational fields affect.

Figure 48: Collision Between Two Black Holes

Illustration: NASA/LIGO

LIGO is the facility that used a laser in a 4km tube to detect the event.

Figure 49: Laser Interferometer
Gravitational-Wave Observatory (LIGO)

LIGO gathered the following data from the observed collision between the two black holes.

Figure 50: LIGO Data Comparison

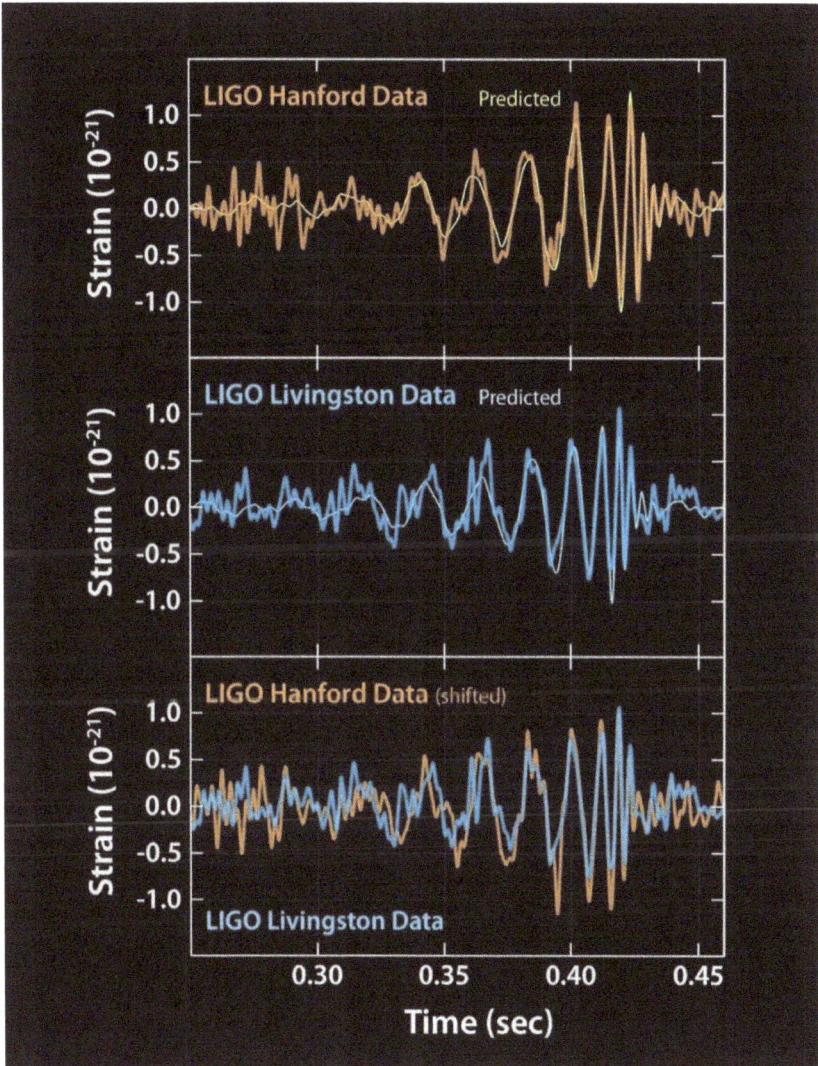

This data and the test does not show whether the effect on the light is as the result of the ripples created by the gravitational force, or the energy created by the collision of the gravitational fields at the collision points. These two are fundamentally different. Gravitational force never ripples. To create a ripple, a constant change must be present; the gravitational force must be variable. This constant change does not exist in gravitational energy. The gravitational force is always constant without any changes. In fact, this steady and never-changing characteristic of gravity is something we heavily depend on and use daily on our planet. A 1kg bar is always 1kg; we always keep a number of samples of it in a safe place in the International Standards building in Paris to make sure 1kg is always 1kg. The weight of the bar comes from the constant pulling force generated by the planet Earth. If the effects of gravity were variable, then the 1kg bar would weigh only half a kilo one day and 2kg the following week. Only a variable gravitational force can create ripples, and because variable gravity does not exist, neither can the ripples.

The ripples are created by the energy released by the colliding fields of gravity. Similarly, magnetic fields are always constant and create a ripple only when they meet another magnetic field. In power generators and electric motors, the magnetic field creates ripples at 50 cycles per second in some countries and 60 cycles per second in others. These ripples are identified by the humans as standard and known as Hertz (50Hz and 60Hz). These variable magnetic fields are created by humans; they don't exist in nature. We also can create variable magnetic fields by using AC power or non-variable magnetic fields by using DC power.

The source of the gravitational force must be of a variable kind in order to create ripples. Some may say the black holes with non-variable gravitational force move at thousands of kilometres per second, come in contact with another black hole, and that will make a variable force that creates the ripples. If this is the case, we should be able to observe a very similar effect between the Earth and the Moon, or the Sun at a lower rate due to lesser gravitational force. The ripples as explained earlier are from the release of energy from the colliding black holes in a very similar way to when two large objects collide and release the energy in the form of a shockwave. The ripples are nothing but shockwaves generated by the colliding black holes. Also, the strong gravitational force can change the direction of streams of energy that exist in the universe, and that change can cause the light to follow the direction of change and appear dimpled.

Gravitation is in the shape of a not-so-perfect sphere, and there is a seamless transition from the strongest form to zero as in the left illustration, rather that staged and sectional form on the right:

Figure 51: Gravitational Forces

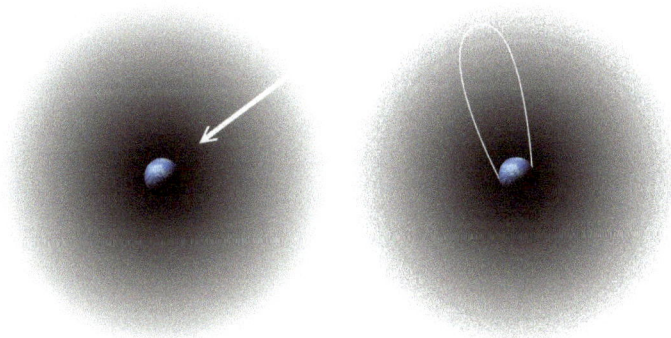

The origin of the gravity is the centre of the entity, and the endpoint is a straight line inward, not a loop back to itself.

The illustration below displays the effects of gravity when two black holes come close together. The gravity increases at the opposite side from where they meet and becomes extremely low to non-existent at the point where they meet and overlap. The gravity fields do not deflect in a way similar to magnetic fields when like poles come close to each other. The non-existent gravity area is created due to the same amount of pulling energy effects at the same time in the same area but in opposite directions. The close proximity of the two large gravitation fields also changes the shape of the field from a perfect sphere to the shape of an egg. The centres of the gravity fields also start moving towards the overlapping area. The overlapping area can disrupt the functioning of the black holes and stop the orbiting entities. This will increase the number of entities present in the area.

We see the effects of gravity of the Moon and the Sun on Earth as a rise in the ocean levels and working of tides.

Figure 52: Gravitational Effect of Two Black Holes Meeting

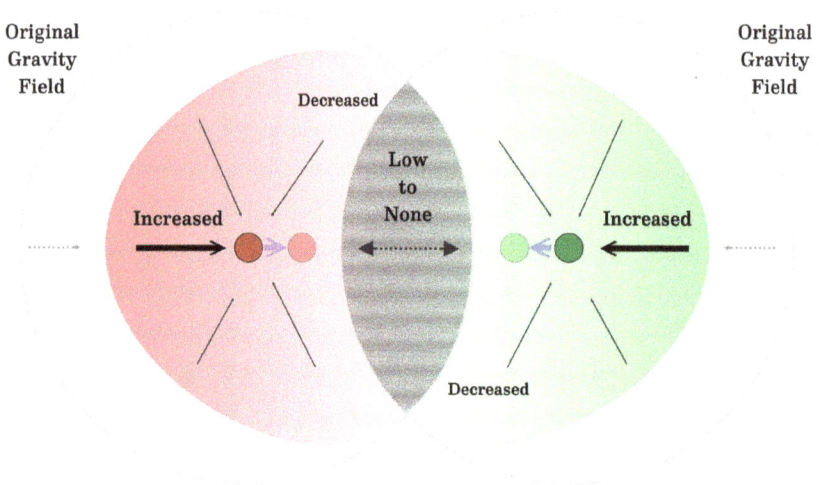

Problems with Understanding the Black Holes

Some hypotheses regarding extra-dimensional space predict micro black holes can be created in particle accelerators such as the LHC (Large Hadron Collider). Concerns have been raised over the result of such tests and the end-of-the-world scenarios. They say, as the result of 2 colliding protons in the accelerator, micro black holes can be created with the lowest amount of energy.

In 1974 Stephen Hawking argued micro black holes can evaporate due to quantum effects. Hawking also renewed his call, in 2015, on the existence of micro black holes and argued the black holes are a passage to another universe. Others also say that the micro black holes can lead to the fourth dimension which is another dimension or another universe.

It is theorised a black hole can be created with a small amount of mass equal or above Plank mass (approx. 22 micrograms). Einstein theorised the photons have a mass (mass-energy equivalent) of $1\times10^{-18}eV/c^2$ ($1eV/c^2 = 1.783\times10^{-36}Kg$).

There are some problems with the concept of these micro black holes, gateway to another dimension, and the fourth dimension. It is simply not happening.

Micro black holes cannot be created due to the lack of at least two main reasons for a black hole's existence, regardless of its size, full-size or micro: absence of extreme gravity and extreme density. Hardly can the combination of any two protons be considered dense—never mind extreme density, and two protons cannot create gravity—never mind extreme gravity!

Black holes in the universe are created at the top of the creation chain of the mother universe. No other influences can overpower or affect their creation and hardly ever their operations. The proposed creation of micro black holes in the LHC is under the extreme influence of electromagnetic energy created by the hadron collider. The influence may be compared, in a way, to a very small piece of magnet, smaller than the tip of a needle, placed in an MRI machine. You have seen images of accidents where large unsecured metal equipment is pulled into the machines. Imagine the power of the small piece of magnet inside the magnetic field of the MRI machine. The magnetic field of the small particle is completely wiped out by the power of the MRI machine.

The LHC is much larger and more powerful than an MRI machine, and the particles are much smaller than the needle tip. We cannot create a micro black hole in a laboratory using magnetic fields. A micro black hole cannot be created on any planet or even close-by any planet or star due to the influence of the much larger body.

Two protons cannot carry enough energy to create a micro black hole. Due to the size of the protons, their masses cannot store enough energy to bridge the energy from the collider and reach the levels required for the creation of a micro black hole. The LHC can speed up the protons close to the speed of light, but the speed is not enough to dwarf the collider's energy and reach the levels for the creation of a black hole. Enough energy cannot be stored in the smashing protons. The energy stored in the protons is used to tear the proton apart, and not enough is left to continue to overcome the power of the collider and create a micro black hole.

The speed of moving protons will reduce from almost the speed of light to zero within a microsecond. If the creation of a black hole requires particles to have the speed of light, or near, then the particles come to the smashing point at the speed of light. At that point, the energy is transferred to the other particles of the opposite side, and forces the particle to break up (Newton's Third Law of Motion page 41, Reaction Vector). The particle pieces come to rest within a centimetre of the collision point; there is just not enough energy left for the particle to travel further than a few centimetres. All the remaining energy released from the collision is neutralised and absorbed by the collider.

The role and the existence of the large magnetic field and its major influence on such small particles cannot be ignored. The black holes in the universe are created free of influence or interference from any external force. These conditions cannot exist in a lab.

One of the characteristics of a black hole is the extreme amount of gravitation force; two protons cannot create a micro black hole with the amount of gravitation energy that exists in a proton. It is not that difficult to calculate the amount of gravitation energy in a proton!!

The gravitational constant is approximately $6.674 \times 10-11$ N·m^2/kg^2 between two bodies. If we use this method to calculate the gravitational energy between the two smashing protons it would be extremely small compared to the extremely large energy injected into the hadron collider. You can see the impossibility of the creation of a micro black hole in the LHC.

The role of the hadron collider is very simple: it creates an extremely narrow electromagnetic tunnel to drive the protons through it. It has two sets of large magnetic coils: one set creates the tunnel, and the second set runs the proton through the tunnel. There are many sets of these coils in the 27km long structure, and every set of accelerators works like a funnel. The receiving entry point is slightly larger than the exit end allowing the proton to enter at slight angle and exit in a straight line and faster.

The amount of electromagnetic energy at the centre of the accelerators is so high that a particle at the speed of light cannot escape its fields. I don't understand how the broken particles after the collision and losing the velocity and speed can create a micro black hole in the middle of a very powerful magnetic field and some of the same particles, of only two protons, can end up in a different universe.

At the centre of full sized black holes, there is nothing much except a lot of slow-moving debris and particles. Slow, because the maximum speed of the black hole is at the outside of the event not in the centre. At the very centre of the event, the effect of gravity is cancelled out by opposing and equal forces from all directions. If there is a way to have a gateway that would lead to some other place, and it is a big IF, then it should be on the outside of the event where the gravity and speed are much higher, not at the centre!!! Wouldn't the gateway to another dimension, if it existed, have already sucked all the particles into the other side and the black hole have disappeared?

Maybe it has disappeared, and because we are receiving the light coming from it billions of years later we still see the images from the past, and if we could see the current images, the black hole would not be there anymore. Or maybe, it was sucked into the gateway and ended up in the other universe and again from the other dimension got sucked back here again and again. This can go forever. Or, what if ...? This fascination will continue.

Now to the fourth dimension. In our natural environment, we can see things in three dimensions. There are One-Dimensional environments (1-D), like a line on paper. Two-Dimensional (2-D), are things like a television screen or computer monitor. The Three-Dimensional environment is the one that we occupy. And the Fourth-Dimension or 4-D which is the other side of the objects in 3-dimensional environments or the back of things we see. We are always able to see only the three sides of every object due to our nature. We are unable to see the fourth sides of objects due to many factors.

My understanding of the fourth dimension comes from what Einstein proposed in early 1900, but now in recent decades it is referred to as another dimension or another universe. A universe parallel to ours—a totally different universe that can co-exist within or side-by-side or parallel to ours.

So, what is this fourth dimension? Is it a gateway to other dimensions? And if it is a gateway, then is the fourth dimension a one-dimensional gateway or is it a three-dimensional passage? If it exists, then how does our dimension work? Do we live in a 3+1-dimensional environment, or if the fourth dimension is a gateway to another 3-D universe then is our environment (3+1) + 3-D? And if we use the gateway and arrive in the new dimension, is it another 3-D environment or is it another 3-D plus 1-D plus 3-D environment ((3+1) + (3+1) +......-D)? You know where this is going! Or maybe the Fourth dimension is a 2-D environment, a flat 1-D and another dimension built in it makes it two-dimensional. So, it is going to be (3+2) + (3+2) +......-D environment. Or is it like ...., (3+2) + (3+2) + (...), ...? Maybe we are located in a dimension before and after dimensions between other dimensions.

This is giving me a bad headache, and it's about to get worse.

Einstein proposed, if we are approaching an object like a box, we are always able to see only three sides of the object: top (or bottom), left and right sides. If our speed increased to the speed of light, we would be able to see the fourth dimension, which is the back of the object, at the same time as the other sides of the fourth dimension. At the speed of light, our three-dimensional environment will become a four-dimensional environment.

The fourth dimension has always existed, but due to our inability we have never been able to observe it at the same time as the other three dimensions. This is due to our physical structure; it doesn't mean it doesn't exist.

We can see objects for two reasons: they emit or reflect light. The emitting light objects are like a torch, a light bulb, the Sun, stars, ... The reflecting light objects are everything that is not a source of light like planets, trees, buildings, ..., etc. The emitting light object can have rays of light coming from all parts and particles of the object, but the reflected light coming from an object can only come within a maximum angle of 179.999....98-degrees. It is a very long number after the decimal point, something like at least 30 decimal points but it is not like 179.999...99, or ends with 99. The decimal points end in ......98. Why not 180°? Why 98 and not 99? Why is such a small decimal point important?

The image below shows the field of view in a 3-dimensional environment. All those lines are at 179.999…98-degree angle:

Figure 53: 3-D Environment Fields of View

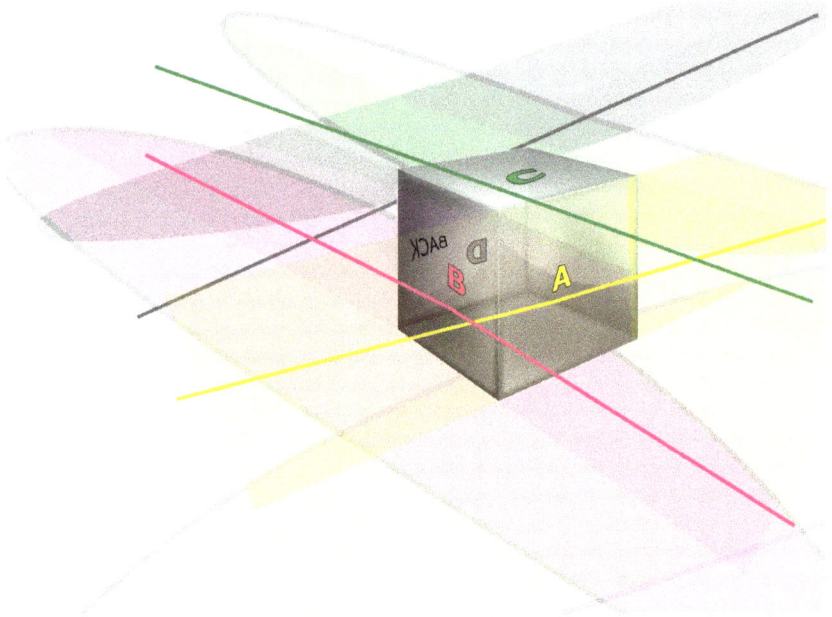

The 3-dimensions are marked as A, B, and C. The fourth dimension is D and is located at the back of the object on the opposite side of A, or can be located on the opposite side of either B or C.

Einstein's theory says we would be able to see A, B, C, and D at the same time if we travel at the speed of light.

The reason we can see objects is the photons make contact with our eyes. If we don't make contact and don't receive these photons, it means we cannot see the object, and there is no light coming from the object.

The surface of the object that the rays of light are coming from is always a flat surface at the point rays are reflected from, or are produced from it. Even though if the object at large is a round shape, the part that the light rays leave is flat at micro, atto (0.000,000,000,000,000,001) or molecular level.

Figure 54: Light Emission/Reflection Angle

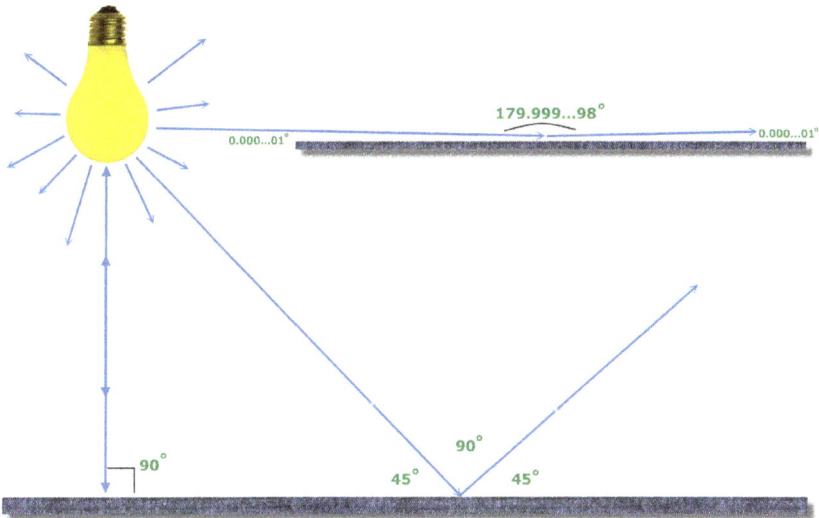

The 0.000...02-degree (or 2 × 0.000...01) is the angle of a ray of light arriving at the point of entry on the object's surface, 0.000...01-degree and the angle that the ray of light is reflected and leaves the surface of the object at 0.000...01-degree. The total angle that a ray of light arrives and leaves is 0.000...02-degree, and it leaves the remaining angle at 179.999...98-degree in order to become a 180-degree angle.

If a ray of light arrives at a 180-degree angle and leaves an object at the same angle is not reflected from the object and is considered parallel to the object and the object cannot be seen or observed. The angle: 0.000...01-degree is the width of the ray of light and is smaller or equal to the width of a photon that makes that ray of light.

If a ray of light is emitted and arrives at the object at a 90° angle, the reflection path is exactly the path that it arrived and it is reflected back to its source. The angle created by this is 0°. The very first and the smallest angle that a ray of light can create is 0.000...01; it arrives at 0.000...01-degree and leaves at 179.999...98° angle. This is the smallest size angle of arrival of light and the widest angle of light possible.

If we closely examine the visibility of an object, we will find objects are only visible within an angle of 0.000...01 and 179.999...98-degrees. Anything greater than that, the object is not visible as the ray of light passes the object in parallel or is reflected from other object's dimensions.

The light always leaves an object at the same angle that it arrived. There is always a fine line at the end of the line of visibility of an object with a width less than the width of one photon.

The width of this line has to be smaller than one photon so only one photon can ever occupy the line, not two photons. One photon of the reflecting dimension. If the line is wide enough to allow the sharing of the boundary line by two photons, then it can be, theoretically, occupied by a photon of each dimension. However, this is not possible because the angle of the light arriving or leaving the object becomes greater than 180° and it is against all laws of physics and the mother universe. The decimal point of 0.000...01 is equal or smaller than the width and size of a photon.

Even if we could travel at the speed of light, we would not be able to see the fourth dimension because, simply, the photons are not there to see. A human brain can process about 24–30 frames of data per second that are streamed to the eyes. Anything more than that is still observed by eye and passed to the brain but the brain is flooded with an overwhelming amount of data and just shuts it down, not processing any until the speed of streaming data to the brain is slowed down and the brain can catch up with it.

If you roll your eyes very quickly from one end to the other end, you will not see the images in between. The only thing you see is the image at the beginning and the image at the end when the eye stops. The images in between are dumped by the brain and not processed. This is due to the slow processing power of the brain.

In order to see the fourth dimension at the same time as the other three, it is not enough just to travel at the speed of light; we also need to be able to see and process at the same rate as the data at the speed of light is reaching us. Otherwise, everything would appear black.

If we ignore this slow processing power of the brain and do not bring it to account but continue to travel at the speed of light towards an object, we will not be able to see the fourth dimension because there are no photons from the fourth dimension to arrive at our eye at the same time as the photons from the other dimensions are arriving. It is the physical absence of photons that is stopping us from seeing all four dimension at the same time.

Even if we double or triple our speed, it still won't make any difference, because not even a single photon can ever exist in the same space and time with other photons from other dimensions. The boundary line is smaller than the width of a photon, and it cannot be shared or widened just because we are speeding.

The rule of 179.999,8 is not something that we have invented; we have only created a method to measure it. This is the rule and mechanics of the universe, and it is the universe that dictates that light cannot be reflected at an angle greater than 179.999...98°. When the viewing angle becomes greater than 179.999...98, the dimension becomes invisible. Theoretically, if we place a viewpoint within 0.000...02° angle, the two dimensions become invisible.

Figure 55: The Invisibility Angle

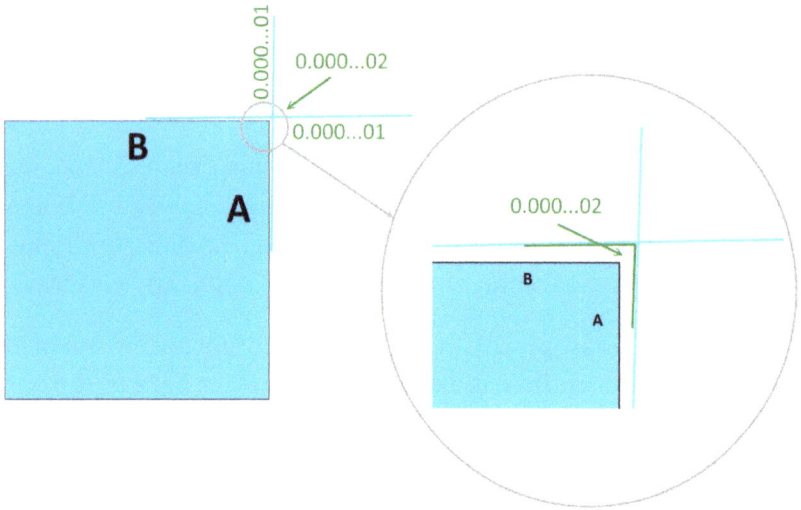

The correct viewing field of an object is between 0.000...01 and 179.999...99-degree that makes the correct viewing angle of 179.999...98-degree, anything beyond this is not visible because the photons don't exist.

There are three scenarios in observing a dimension: 1—Travelling at normal speed and processing at normal speed. 2—Travelling at the speed of light and processing at normal speed. 3—Travelling at the speed of light and processing at the speed of light. If we approach an object at normal speed, we can see the one dimension up to 179.999...98-degree and then nothing of that dimension. There is a viewpoint of 180.000...02-degree between the two A and D dimensions.

Figure 56: The Visibility Angle

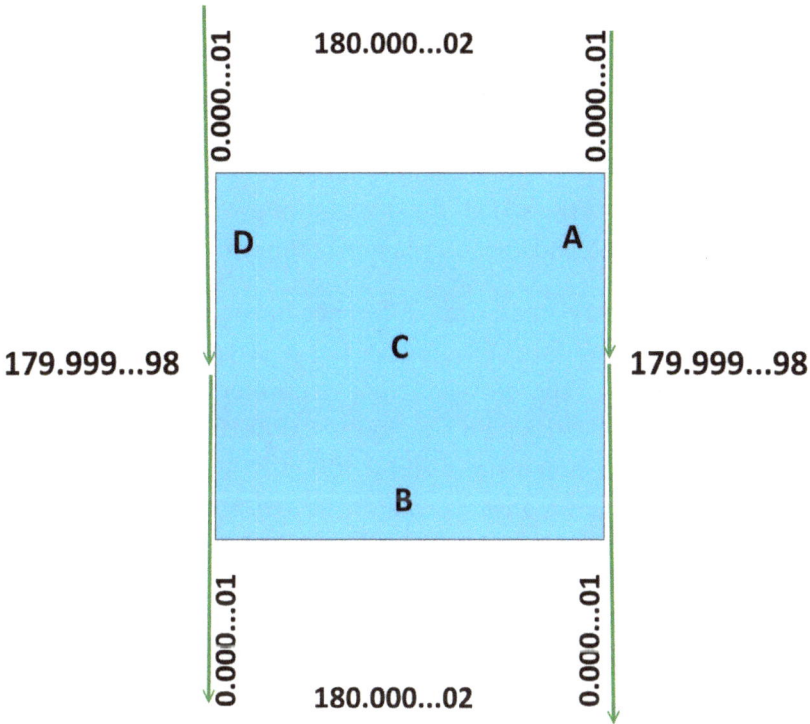

180.000...02

0.000...01

0.000...01

D

A

C

179.999...98

179.999...98

B

0.000...01

0.000...01

180.000...02

If we are processing this at normal speed, we will not be able to see the two dimensions at the same time due to the absence of photons. If we are approaching it at the speed of light and processing it at normal speed, then we see all black due to our inability to process fast enough.

If we are approaching an object at the speed of light and processing at the same speed, then we can still see the two dimensions A and D in the same way as normal. We are not able to see the fourth dimension at the same time as dimension A. The process will be exactly same as the first instance but at much faster speed.

Maybe it is not so much about photons, seeing and visibility. Being present in all four dimensions at the same time is what Einstein was talking about. Seeing or not, blindfolded or not; dimensions always end at 180-degrees, accept it or not. The boundary of a dimension is between 0° and 180°, beyond this, we are in a different dimension.

We cannot be present in dimensions that are not in the same time and space. Travelling at the speed of light can put us in one dimension at one fraction of time, and another fraction of time in another dimension.

There will be a very small time difference between our presence in either dimension and that fraction is what separates the dimensions. Einstein's theory of space-time cannot remove or bridge the separation fraction. The separation fraction is 0.000...02-degree between the two adjoining dimensions and 0.000...02° + 180° between the two non-adjoining dimensions, and the time differential is the minimum time it takes to bridge this angle regardless of the speed.

Here comes another headache:

The Fifth Dimension's Separation Fraction in a five-dimensional object is:

0.000...02 + B + D because the fifth dimension (E) is located next to the fourth dimension (D) the minimum distance is the length of B (180°) and length of D (180°). There is the first separation fraction of 0.000...02 between A and B dimensions, and 0.000...02 between D and E dimensions, and 180 between A and D (or total length of B).

The fraction is variable, and it directly depends on the size of the object. It can be a fraction of a second if the object is a small box, or millions of light years if the object is a galaxy. Regardless of size, it is a distance between dimensions, and it requires time to cross. This is what stops us to being in two places at one time even if we travel at the speed of light; it is the light that creates this fraction in the first place.

Maybe the whole fourth dimension is not about objects, and it is all about the fourth dimension of the universe. More accurately, it is about the universe parallel to ours. It is a whole different story if the reference to the fourth dimension is intended to be a different universe. If this is the case, then the universe is only a one-dimensional environment on the inside and it is the dimension we are in it! The inner-dimension.

Imagine a very large multi-dimensional environment and that we are in it. Not only us, but the whole universe and everything that ever exists is inside this environment, so how many dimensions can there be? If you are inside a sphere how many dimensions can you have? Only one, the one all around.

Sure there are a lot of three-dimensional objects inside, but the universe itself is one-dimension in the inside. If we look at this universe from outside, somewhere in the greater universe trillions of light years away from the fringes, then we would be able to see the dimensions of this universe; otherwise looking inside out we can see only one dimension, the inner-dimension.

Interestingly, this multi-dimensional universe is located inside the greater universe in a similar way to the multi-dimensional objects in this universe. Inside a multi-dimensional environment, it is always a straight line between the dimensions. The closest thing that comes to mind is something like a Matryoshka doll. An environment, inside an environment, inside a galaxy, inside a cluster. Inside a super-cluster. Inside a universe. Inside a universe. Confused?

The Classification of the Creation

Two different methods can be used to classify the creation: Stage, and Tier. The stage classification of the creation is about the segmented events that occur during the process of creation from the beginning of the creation of the universe to the complete end of it. These stages recur within the tier and can start in one tier and end in another tier. The stages can be divided to sub-stages to cover shorter events.

The Tiers are about the particles, the creation of particles, and their transformation from one form to a new form of particles. What happens during this transition is covered under Stages. An event can begin in one stage and end in another stage, but the tier classification is based on the beginning and the ending of an event not the details of it. The length, start and end of a tier is predictable, but not the stages as the stages can change direction or end prematurely due to the nature of events. As you can see below, the tiers can be predicted into the far distant future but not the stages. We cannot predict stages beyond the current time; we can guess not predict. There also can be gaps between the end of one stage and the start of another. Tiers are connected back-to-back, and the end of one tier is the start of the next.

The creation is classified into seven tiers, and, so far, three stages:

TIER	FROM CREATION OF	TO CREATION OF
0	Infinity — Non-existence of any kind	Type-0 Particles — Creation Begins
1	Type-0 Particles — Universe Begins	Quarks and Dark Matter
2	Quarks	Subatomic and Molecules — End of creation of particles
3	Molecules	The Big Event, beginning of Life
4	Life	End of Creation— End of current existence
5	End of Existence	Death of last star
6	Last Star	Complete destruction of current universe—End of Universe

Events in green and red are under Stages classification.

Every tier begins with the creation of new particles and continues to the transformation of that particle to a new particle. From the end of tier 6, the creation is not going back to tier 1 or tier 0 as the creation does not work like a circle of life and recycle itself. Decayed particles cannot be re-created into new forms of particles of any kind.

The size of tiers below indicates amount of activities in this tier:

Figure 57: The Tiers and Stages of Creation

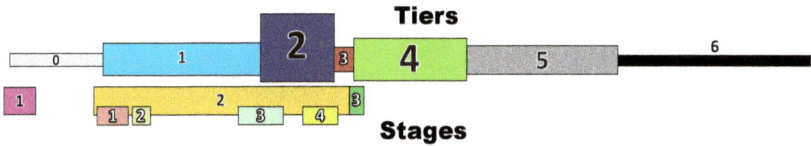

As the image illustrates, the beginning of Stages and Tiers do not meet, stages always start earlier at the starting point of the event, but the tiers start at the final result of the event. If an event results in the creation of subatomic particles and the event takes 3 5 trillion years to complete, the beginning of the event is the stage's starting point and the time when the first batch of subatomic particles are created is the tier's starting point. This difference of starting points can be from hundreds of thousands to hundreds of millions of years.

Why the Big Bang?

In 1927 Georges Lemaître, an astrophysicist, suggested the expanding universe could be traced back in time to a singular starting point of expansion. Later, in 1968 and 1970 papers were published by three British astrophysics: Stephen Hawking, George F. R. Ellis, and Roger Penrose. Their calculations extended to the theory of General Relativity, showed time and space had a definable beginning, and that it corresponded with the origin of matter and energy.

Figure 58: Divergence Over Time from the Singularity

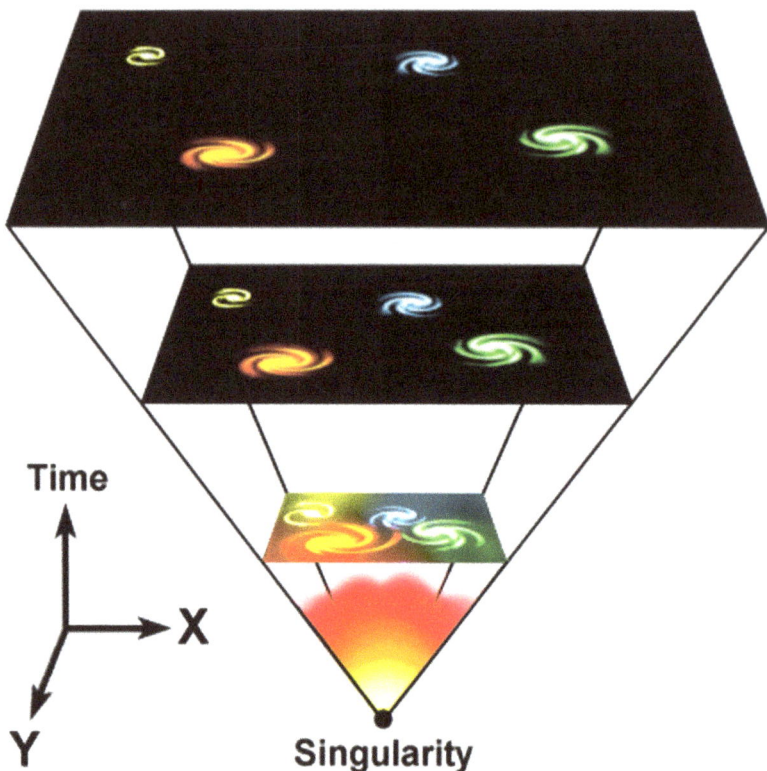

Redshift is when the wavelength of a signal and light beam increases and is shifted towards the red end of its spectrum. When a beam of light is passed through a prism and the red becomes redder and the wavelength of the beam shifts towards red it indicates the beam of light is coming from a source that is moving away from the observer. When the wavelength decreases and shifts away from the red, it indicates that the source of the light is moving towards the observer. This method is used to track sources of light, like stars from the far distance, to determine whether the entity is moving away from, or towards, our planet. The beam of collected light is tested against a beam created in the lab to compare the shift from the collected light to a fresh and un-shifted sample. Blueshift is the opposite of redshift and referred to as negative redshift; by observing a beam of light, the shift towards the blue means the source is moving closer.

The redshift test is very similar to the test of sound. When a source of sound is moving towards the listener, the frequency heard increases, and the pitch is higher. When it is moving away, the frequency is reduced, and the pitch is lower. The redshift method is used to confirm the expansion of the entities, and their movement away can be traced back to a singular event and a common centre.

Redshift is not a reliable tool for this task due to its predictability, lack of integrity of the source, and compromised and contaminated light fragments. Redshift from some sources are predictable, and the amount of shift can be calculated. Redshift from some other sources cannot be trusted due to the quality and the integrity of the source and the integrity of the beam of light that has sometimes travelled for 12 billion years. There is an uncertainty to all the tests and the source and the subject of the test; and although there are some sources and test subjects that can be trusted, the large percentage of unreliable sources makes it difficult to trust the result and the accuracy of the test. The further distant the sources are, the more unreliable they become.

Figure 59: The Effect of Redshift

Unfortunately, the Big Bang is a bad title for this event, as it doesn't sit very well with what really happened. The 'bang' is the wrong term to describe such an event. Bang indicates 'sudden' and 'loud', and a one-off event. The event wasn't sudden, nor was it a one-off: it was a series of events and took millions of years to complete, and there were hardly any bangs in it. If the event were a one-off explosion that suddenly erupted, it would have wiped the epicentre very clean and created a very large void. It also needed an extreme amount of energy to create such a bang so suddenly. At that point of the creation, the universe did not have enough resources to provide and supply such an amount of energy. The area of the universe needed for the energy to explode was millions of cubic light-years across and a source to reach all this area at once, and cover every square inch to create such a bang was not available to the creation.

The sheer number of extremely large entities, and the vast area of the universe, makes it impossible for the creation to overcome the demand for the required energy. Not that the universe has this amount of energy at its disposal even now after everything is created and a larger number of complex molecules are present. Although energy was created at the time of the big event, and a lot of it was stored in a great number of entities, still, such a vast area and large number of entities needed an extreme amount of energy to make it possible for the sudden event. That energy just wasn't there. Looking back at the events, it wasn't a bang at all, but it was bigger than big, that started the chain of events. The event was more like a series of bangs over millions of years allowing the physical expansion of the area ("physical expansion of the area", NOT the "physical expansion of the universe", the universe never expands, and it always remains an infinite size at all times. See Chapter 7). The Big Event started an expansion of the extremely concentrated entities back to what they were before the time they became so dense and concentrated.

There was one single explosion that started the whole event, and it wasn't responsible for the big bang. Billions of years earlier, there had been even larger events that would dwarf the big bang; however, due to the absence of enough entities, these events did not lead to the larger event that created the universe we know now. Once the universe had enough energy to create such a bang there were not enough entities present to go full scale; and when the universe had the required number of entities, there wasn't enough energy to create the bang. Therefore, the creation had to go slow on the bang and have many smaller bangs rather than one big one. The events that took place, from leading to the big event to the completion of it, were more like a domino effect, and a correlated chain of events, rather than a one-off event. The Big Event is a more suitable title for Phase 4 of creation than the big bang.

Although time started more than a trillion years earlier, the Big Event was a disruption in the timekeeping process. Time continued during the Big Event, but it recorded a different series of events, and those events either do not exist, or connect the previous events to the current one, and, therefore, it cannot be used as the Time we need to use. Time never stopped, but the Big Event interrupted it, and all tangible evidence that could connect us to the time before the Big Event has been destroyed. UniTime continues to function as usual but the local times have been reset and restarted. The large majority of local times will not start for billions of years due to the activities of the Big Event, but they did as soon as the activities ended. The UniTime can be traced back to before the Big Event if we have the right tools, and it will be a continuous time to register the events after the Big Event to the present day.

SOURCES AND REFERENCE MATERIAL

The Boomerang Nebula

The Boomerang Nebula is a young planetary nebula and the coldest entity found in the Universe so far. The NASA/ESA Hubble Space Telescope image is yet another example of how Hubble's sharp eye reveals surprising details in celestial entities.

This NASA/ESA Hubble Space Telescope image shows a young planetary nebula known (rather curiously) as the Boomerang Nebula. It is in the constellation of Centaurus, 5,000 light-years from Earth. Planetary nebulae form around a bright, central star when it expels gas in the last stages of its life.

The Boomerang Nebula is one of the Universe's peculiar places. In 1995, using the 15-metre Swedish ESO Submillimetre Telescope in Chile, astronomers Sahai and Nyman revealed that it is the coldest place in the Universe found so far. With a temperature of −272C, it is only 1 degree warmer than absolute zero (the lowest limit for all temperatures). Even the −270C background glow from the Big Bang is warmer than this nebula. It is the only entity found so far that has a temperature lower than the background radiation.

Keith Taylor and Mike Scarrott called it the Boomerang Nebula in 1980 after observing it with a large ground-based telescope in Australia. Unable to see the detail that only Hubble can reveal, the astronomers saw merely a slight asymmetry in the nebula's lobes suggesting a curved shape like a boomerang. The high-resolution Hubble images indicate that 'the Bow Tie Nebula' would perhaps have been a better name.

The Hubble Telescope took this image in 1998. It shows faint arcs and ghostly filaments embedded within the diffuse gas of the nebula's smooth 'bow tie' lobes. The diffuse bow-tie shape of this nebula makes it quite different from other observed planetary nebulae, which normally have lobes that look more like 'bubbles' blown in the gas. However, the Boomerang Nebula is so young that it may not have had time to develop these structures. Why planetary nebulae have so many different shapes is still a mystery.

The general bow-tie shape of the Boomerang appears to have been created by a very fierce 500,000 kilometre-per-hour wind blowing ultra-cold gas away from the dying central star. The star has been losing as much as one-thousandth of a solar mass of material per year for 1,500 years. This is 10–100 times more than in other similar entities. The rapid expansion of the nebula has enabled it to become the coldest known region in the Universe.

The image was exposed for 1000 seconds through a green-yellow filter. The light in the image comes from starlight from the central star reflected by dust particles.

Date: 20 February 2003
Source: European Space Agency, NASA
spacetelescope.org

Planck Epoch

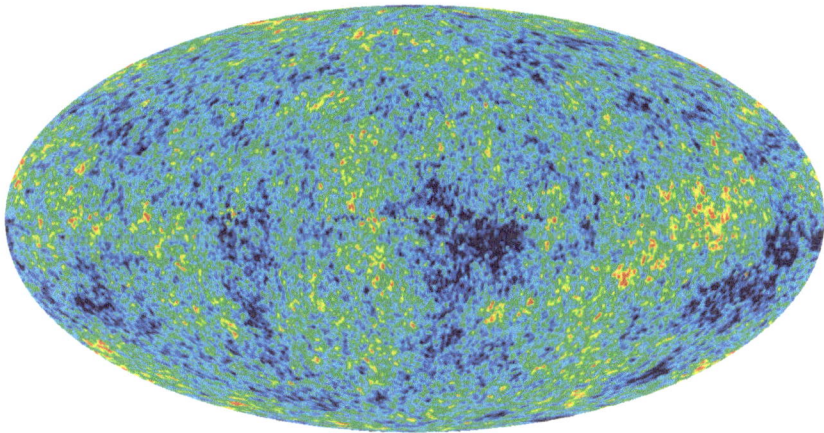

Source: Wikipedia

The Planck epoch (named after Max Plank 1858 – 1947 a German Physicist) is an era in traditional (non-inflationary) Big Bang cosmology in which the temperature is high enough that the four fundamental forces—electromagnetism, gravitation, weak nuclear interaction, and strong nuclear interaction—are all unified in one fundamental force. Little is understood about physics at this temperature, and different theories propose different scenarios. Traditional Big Bang cosmology predicts a gravitational singularity before this time, but this theory is based on general relativity and is expected to break down due to quantum effects. Physicists hope that proposed theories of quantum gravitation, such as string theory, loop quantum gravity, and causal sets, will eventually lead to a better understanding of this epoch. In inflationary cosmology, times before the end of inflation (roughly 10^{-42} seconds after the Big Bang) do not follow the traditional Big Bang timeline. The universe before the end of inflation is a near-vacuum with a very low temperature and persists for much longer than 10^{-32} second. Times from the end of inflation are based on the Big Bang time of the non-inflationary Big Bang model, not on the actual age of the universe at that time, which cannot be determined in inflationary cosmology. Thus, in inflationary cosmology, there is no Planck epoch in the traditional sense, though similar conditions may have prevailed in a pre-inflationary era of the universe.

A new, dynamic portrait of our Milky Way galaxy shows a frenzy of gas, charged particles and dust. The main composite image comes from the European Space Agency's Planck mission, in which NASA plays an important role. It is constructed from observations made at microwave and millimetre wavelengths of light, which are longer than what we see with our eyes. The various components making up the main image are shown below it: Dust Glow (upper left). The red colours making up this map show light coming from the thermal glow of dust throughout our galaxy. The dust is cold, only about 20 degrees above absolute zero (20 Kelvin). Carbon Monoxide Gas (upper right). Yellow shows carbon monoxide gas, which is concentrated along the plane of our Milky Way in the densest clouds of gas and dust that are churning out new stars. Careening Particles (lower left). The green shows a kind of radiation known as free-free. This occurs when isolated electrons and protons careen past one another in a series of near-collisions, slowing down but continuing on their own way (the name free-free comes from the fact that the particles start out alone and end up alone). The free-free signatures are associated with hot, ionised gas near massive stars. Captured in Magnetic Fields (lower right). Blue indicates a type of radiation called synchrotron, which occurs when fast-moving electrons, spit out of supernovas and other energetic phenomena, are captured in the galaxy's magnetic field, spiralling along them near the speed of light.

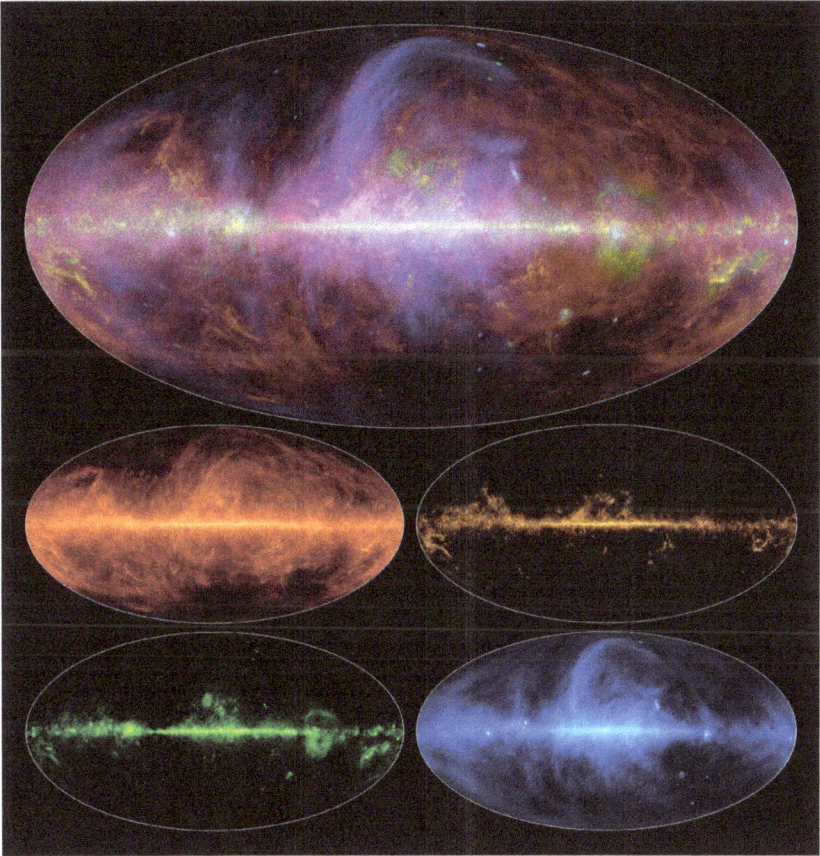

Credit: ESA/NASA/JPL-Caltech

Quark Structure of a Proton

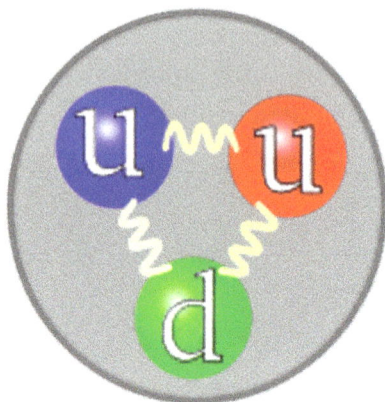

Original artwork: Arpad Horvath | Source: Wikipedia

CERN – CMS Experiment

© 2008 CERN – CMS detector | Photograph: Hoch, Michael; Brice, Maximilien |
Date: 01/08/2008

Periodic Table of Chemical Elements

Source: Wikipedia | Author: User: Cepheus

Night Sky and Southern Cross

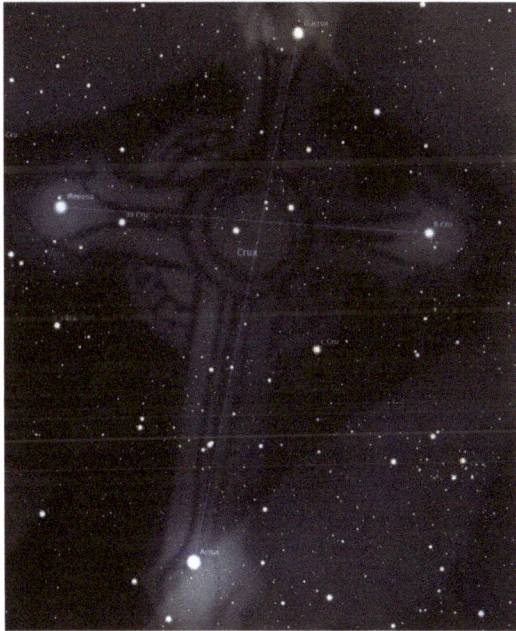

These pictures were generated from live data and real time using open source software:

Source: Stellarium http://www.stellarium.org

Theory of Relativity

The theory of relativity, or simply relativity in physics, usually encompasses two theories by Albert Einstein: special relativity and general relativity.

Concepts introduced by the theories of relativity include:

Measurements of various quantities are relative to the velocities of observers. In particular, space contracts and time dilates.

Spacetime: space and time should be considered together and in relation to each other.

The speed of light is nonetheless invariant, the same for all observers.

In physics, Spacetime (also space–time, space time or space–time continuum) is any mathematical model that combines space and time into a single interwoven continuum. The Spacetime of our universe is usually interpreted from a Euclidean space perspective, which regards space as consisting of three dimensions, and time as consisting of one dimension, the "fourth dimension". By combining space and time into a single manifold called Minkowski space, physicists have significantly simplified a large number of physical theories, as well as described in a more uniform way the workings of the universe at both the super-galactic and subatomic levels.

Source: Wikipedia

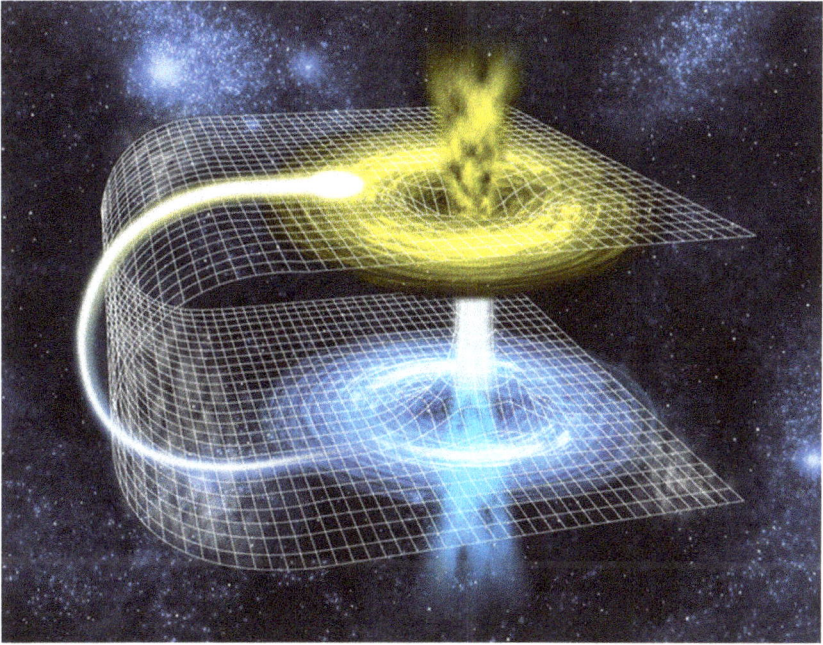

Space-Time Experiment

May 4, 2011: Einstein was right again. There is a space-time vortex around Earth, and its shape precisely matches the predictions of Einstein's theory of gravity.

Researchers confirmed these points at a press conference today at NASA headquarters where they announced the long-awaited results of Gravity Probe B (GP-B).

"The space-time around Earth appears to be distorted just as general relativity predicts," says Stanford University physicist Francis Everitt, principal investigator of the Gravity Probe B mission.

Source: NASA
Author: Dr Tony Phillips
Date: May 2011

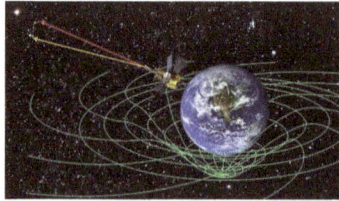

Source: Stanford University | Author: James Overduin, Jan 2008

Space-Time Deflection

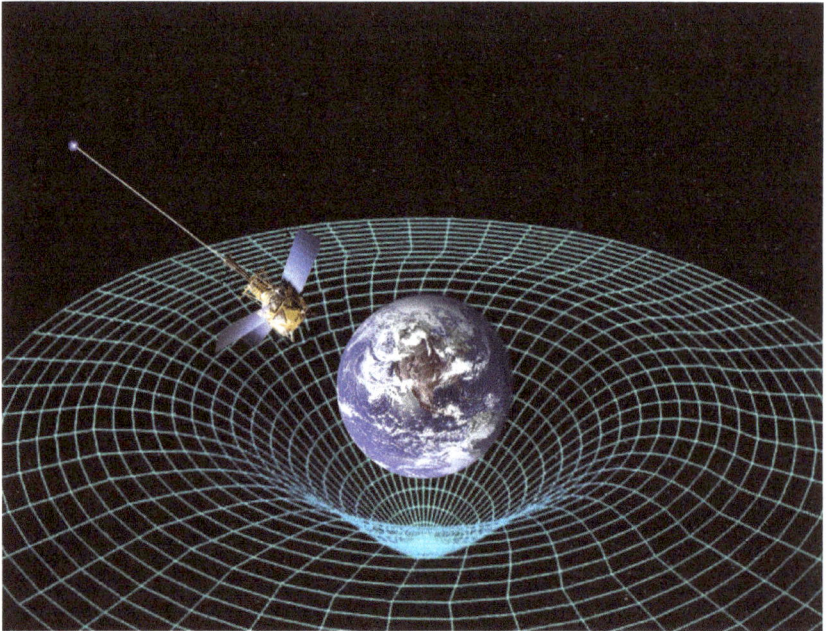

Source: NASA | http://www.nasa.gov/mission_pages/gpb/gpb_012.html |
Art Work: Original work of NASA

Artist concept of Gravity Probe B orbiting the Earth to measure space-time, a four-dimensional description of the universe including height, width, length, and time.

Date: 18 May 2008

Hafele–Keating Experiment

The Hafele–Keating experiment was a test of the theory of relativity. In October 1971, Joseph C. Hafele, a physicist, and Richard E. Keating, an astronomer, took four cesium-beam atomic clocks aboard commercial airliners. They flew twice around the world, first eastward, then westward, and compared the clocks against others that remained at the United States Naval Observatory. When reunited, the three sets of clocks were found to disagree with one another, and their differences were consistent with the predictions of special and general relativity.

The measured change in rate was $(29\pm1.5) \times10^{-14}$, consistent with the result of 30.7×10^{-14} predicted by general relativity.

Source: Wikipedia

Black Holes

A black hole is a mathematically defined region of Spacetime exhibiting such a strong gravitational pull that no particle or electromagnetic radiation can escape from it. The theory of general relativity predicts that a sufficiently compact mass can deform Spacetime to form a black hole. The boundary of the region from which no escape is possible is called the event horizon. Although crossing the event horizon has an enormous effect on the fate of the entity crossing it, it appears to have no locally detectable features. In many ways, a black hole acts like an ideal black body, as it reflects no light. Moreover, quantum field theory in curved Spacetime predicts that event horizons emit Hawking radiation, with the same spectrum as a black body of a temperature inversely proportional to its mass. This temperature is on the order of billionths of a kelvin for black holes of stellar mass, making it essentially impossible to observe.

Source: Wikipedia

Dust Disc Around a Black Hole in Galaxy NGC 7052

Source: Hubble Telescope | Credit: Roeland P. van der Marel (STScI),
Frank C. van den Bosch (Univ. of Washington), and NASA

Time Dilation

Time dilation is a difference in elapsed time between two events as measured by observers either moving relative to each other or differently situated from gravitational masses.

Space Contraction is the phenomenon of a decrease in length measured by the observer, of an entity which is travelling at any non-zero velocity relative to the observer.

Space and Time should be considered together and in relation to each other.

The speed of light is constant to all observers.

Source: Wikipedia

Central Parts of the Milky Way

VISTA gigapixel mosaic of the central parts of the Milky Way | Date: 2012 |
Credit: ESO/VVV Survey/D. Minniti |
Acknowledgement: Ignacio Toledo, Martin Kornmesser

Space Shuttle

The Space Shuttle is the world's first reusable spacecraft and the first spacecraft in history that can carry large satellites both to and from orbit. The Shuttle launches like a rocket, manoeuvres in Earth orbit like a spacecraft and lands like an aeroplane. Each of the three Space Shuttle orbiters—Discovery, Atlantis and Endeavour—is designed to fly at least 100 missions. So far, altogether they have flown a combined total of less than one-fourth of that.

Specifications of *Endeavour* (OV-105)

Length: 37.237 m

Wingspan: 23.79 m

Height: 17.86 m

Empty weight: 78,000 kg

Gross lift-off weight: 110,000 kg

Maximum landing weight: 100,000 kg

Maximum payload: 25,060 kg

Payload to low Earth orbit: 24,310 kg

Payload bay dimensions: 4.6 × 18 m

Operational orbit altitude: 190 to 960 km

Orbital speed: 7,743 m/s (27,870 km/h)

Crew: minimum of two, typically seven, maximum 11z

Source: ESA, NASA

Universe Expansion—and the Big Bang

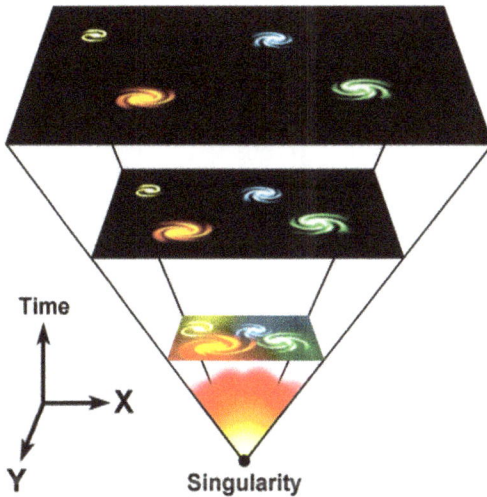

This illustration shows abstracted "slices" of space at different points in time. It is simplified as it shows only two of three spatial dimensions, to allow for the time axis to be displayed conveniently.

Date: Oct 2004
Author: User: Fredrik
Source: Wikipedia

The Big Bang theory is the prevailing cosmological model of the universe from the earliest known periods through its subsequent large-scale evolution. It states that the universe expanded from a very high-density state, and offers a comprehensive explanation for a broad range of observed phenomena, including the abundance of light elements, the cosmic microwave background, large scale structure, and Hubble's Law. If the known laws of physics are extrapolated beyond where they are valid, there is a singularity. Modern measurements place this moment at approximately 13.8 billion years ago, which is thus considered the age of the universe. After the initial expansion, the universe cooled sufficiently to allow the formation of subatomic particles, and later simple atoms. Giant clouds of these primordial elements later coalesced through gravity to form stars and galaxies.

In the mid-20th century, three British astrophysicists, Stephen Hawking, George F. R. Ellis, and Roger Penrose turned their attention to the theory of relativity and its implications regarding our notions of time. In 1968 and 1970, they published papers in which they extended Einstein's theory of general relativity to include measurements of time and space. According to their calculations, time and space had a finite beginning that corresponded to the origin of matter and energy.

Source: Wikipedia
Date Retrieved: 2012

Dust Storm: Mungerannie, South Australia

Date: Jan 2010 | Author: Sydney Oats | Source: Flicker, Public Domain | https://www.flickr.com/photos/57768042@N00/4399878827

Effect of Gravity

Images from Wikipedia, space.com, and NASA, and other science sites

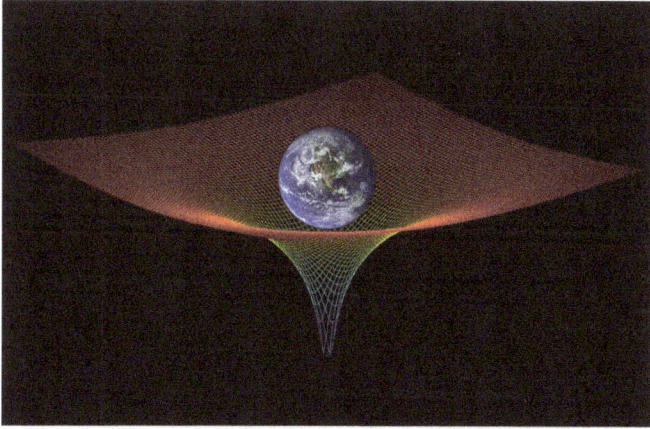

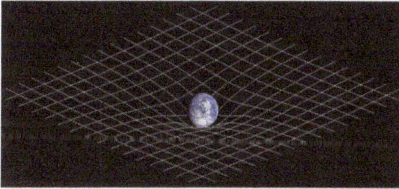

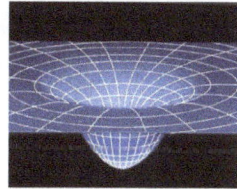

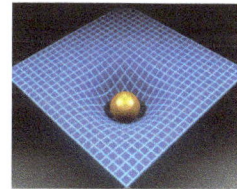

Source: Public Domain, Credits to their creators.

Super Massive Stars

Nine monster stars with masses over 100 times the mass of the sun in the star cluster R136

The image shows the central region of the Tarantula Nebula in the Large Magellanic Cloud. The young and dense star cluster R136 can be seen at the lower right of the image. This cluster contains hundreds of young blue stars, among them the most massive star detected in the Universe so far. Using the NASA/ESA Hubble Space Telescope astronomers were able to study the central and most dense region of this cluster in detail. Here they found nine stars with more than 100 solar masses.

Credits: NASA, ESA, P Crowther (University of Sheffield)

The Carina Nebula: Star Birth in the Extreme

Credit: NASA, ESA, N. Smith (University of California, Berkeley),
and the Hubble Heritage Team (STScI/AURA)

Star Cluster NGC 290

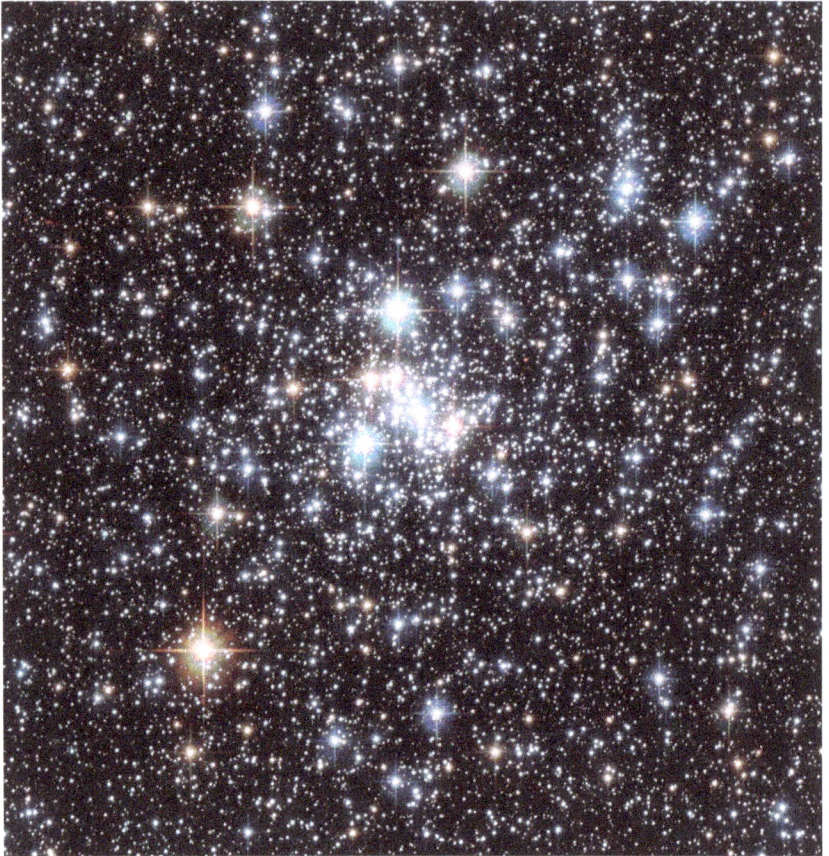

Credit: European Space Agency & NASA
Acknowledgement: E. Olszewski (University of Arizona)

The Tarantula Nebula in the Large Magellanic Cloud

Credit: NASA, ESA, ESO, D. Lennon (ESA/STScI),
and the Hubble Heritage Team (STScI/AURA)

Spiral Galaxy M81

Credit: NASA, ESA, and the Hubble Heritage Team (STScI/AURA)

Carina Nebula: Great Clouds NGC 3372

Credit: NASA, ESA, N. Smith (University of California, Berkeley),
and the Hubble Heritage Team (STScI/AURA)

Interacting Galaxy Group, Stephan's Quintet (HCG 92)

Credit: NASA, ESA, and the Hubble SM4 ERO Team

Interacting Spiral Galaxies NGC 2207 and IC 2163

Credit: NASA, ESA, and the Hubble Heritage Team (STScI)

Globular Star Cluster Omega Centauri

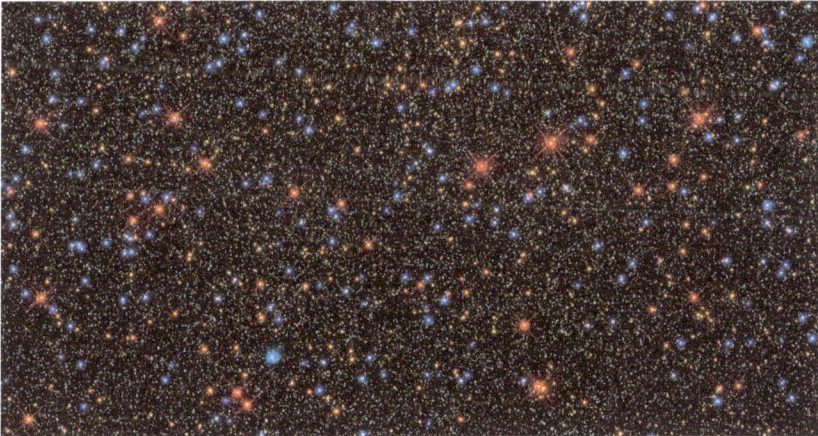

Credit: NASA, ESA, and the Hubble SM4 ERO Team

Hubble Ultra-deep Field: Constellation Fornax

Credit: NASA and the European Space Agency